W0254029

PROTOPLASMATOLOGIA

HANDBUCH DER PROTOPLASMAFORSCHUNG

HERAUSGEGEBEN VON

L. V. HEILBRUNN UND F. WEBER
PHILADELPHIA GRAZ

BAND VIII

PHYSIOLOGIE DES PROTOPLASMAS

6

FROST, DROUGHT, AND HEAT RESISTANCE

WIEN
SPRINGER-VERLAG
1958

FROST, DROUGHT, AND HEAT RESISTANCE

BY

J. LEVITT
COLUMBIA/MISSOURI

WITH 29 FIGURES

WIEN
SPRINGER-VERLAG
1958

ISBN-13: 978-3-211-80490-2 e-ISBN-13: 978-3-7091-5463-2
DOI: 10.1007/978-3-7091-5463-2

Protoplasmatologia
VIII. Physiologie des Protoplasmas
6. Frost, Drought, and Heat Resistance

Frost, Drought, and Heat Resistance

By

J. Levitt

University of Missouri, Columbia, Missouri, U. S. A.

With 29 Figures

Contents

Introduction

With the exception of Chapter 5, the present work deals solely with plants. This was necessary because the author is a botanist and not familiar with the animal literature. It is also reasonable, since tolerance plays a greater role in the resistance of plants. The greater complexity of animal morphology, which may lead to the death of a whole organism as a result of injury to a single organ, also results in basic differences between the kinds of resistance in the two groups. In recent years, however, much work has been done on the frost resistance of individual animal cells, and this is more strictly comparable to the plant work. These researches prompted the author to include a chapter on frost resistance in animals, in the hope that this work may help to solve the whole problem for both plants and animals.

The recent Russian literature is briefly reviewed by KURSANOW (1956) but cannot be analyzed or interpreted by the author in the absence of the original papers.

1. Terminology

One pitfall the scientist must learn to avoid is a controversy of a purely semantic nature. This can lead to an enormous waste of time and energy. It can best be avoided by the adoption of exactly defined terms; for clear concepts require clear terminology. There is nowhere a greater need for this than in the field of frost, drought, and heat resistance. A good example is STOCKER's (1956) objection to LEVITT's (1951) statement that resistance to one carries with it a resistance to the other two. The objection is valid if resistance is used in a general sense, invalid if used in MAXIMOV's (1929) sense, as it was intended to be. Yet STOCKER himself was referring specifically to his so-called protoplasmic resistance. Such differences in usage convinced the author that the terminology in this field needs a complete overhauling. This does not mean that old, established terms should be discarded, but that they should be defined more precisely. New terms should be kept to a minimum.

The plant may conceivably develop a resistance not only to the above three injurious environmental factors, but to others as well. The general phenomenon should therefore be called *environment resistance,* which may be defined as the ability of an organism to survive an unfavorable environment. The term is analogous to disease resistance, and it implies resistance to a specific environmental factor. BIEBL (1952) has used the term "ecological resistance" in the same sense, but this sounds analogous to physiological resistance, i. e., it could mean the ecological resistance of a plant to infectious disease. The same objection holds against the use of the general term "physiologic resistance" for phenomena such as frost, drought, and heat resistance. Environment resistance, on the other hand, has no relation to infectious diseases. It simply implies the ability to survive a harsh environment.

But is this definition complete? Should resistance be confined to survival, or should it also include the ability to grow under harsh environmental conditions? It may not always be possible to separate the two, since (a) injured tissues cannot grow, so a reduction in growth is certain to accompany injury, and (b) a prolonged growth stoppage may ultimately prevent survival. STOCKER (1956), in fact, considers drought injury to have the same cause as growth reduction—a direct effect of drought on protoplasmic structure, which at first affects growth, but when more pronounced produces injury. A case in point is the ability of a winter wheat variety to survive the winter. According to HELMERICK and PFEIFER (1954), this may depend on its ability to germinate under conditions of low moisture supply and to produce enough growth before frost sets in.

Ability to grow under conditions of low moisture may certainly be a factor in drought survival. SATOO (1956) found that *Pinus densiflora* was the most resistant of the three species tested and its growth was decreased only when the soil moisture was depleted below permanent wilting. The growth of the less drought resistant *Cryptomeria japonica* and *Chamaecyparis obtusa*, on the other hand, was affected by a slight decrease of soil moisture below field capacity. In accord with the WALTER (1956) school of thought, BAUMAN (1957) followed the effects of water deficiency

Table 1. *Relation between osmotic value (atms.) and yield of irrigated and unirrigated plants.*
(From BAUMAN 1957.)

Oats				Sugar beets			
Water supplied (in.)	Water used (in.)	Av. osm. val. (atms.)	Yield (fresh wt.)	Water supplied (in.)	Water used (in.)	Av. osm. val. (atms.)	Yield (fresh wt.)
0	10.9	14.4	62.8	0	10.0	22.4	5.3
4.0	15.4	10.1	77.9	8.0	15.7	12.6	15.0
6.5	15.7	10.6	71.1	7.0	17.3	12.1	15.9
5.5	15.3	10.4	67.7	10.0	21.3	13.6	12.8
4.5	15.4	11.0	66.5	8.5	18.1	13.6	13.4
8.5	19.2	9.4	95.2	11.0	19.0	13.2	12.7

by determining the osmotic value (or "hydrature") of the plant. As had been previously found by BERNSTEIN and PEARSON (1954), yield was related inversely to osmotic value (Table 1) though neither the quantity of water added nor the amount used gave a good correlation with yield. He distinguished five phases of "hydrature":

A. Optimum hydrature. Normal water supply and growth, no added water needed. Must be maintained for largest yield.

B. Sligthly reduced hydrature. Growth decreases day to day. Water requirement increases in proportion to the distance of the osmotic value from the optimum.

C. Greater hydrature decrease. Growth almost completely stopped.

D. Hydrature at which growth stops. No harvest can be expected.

E. Lethal hydrature.

This would seem to agree with STOCKER, since injury is apparently the last stage of growth reduction. But is must be realized that drought resistance and yield are not necessarily correlated. As pointed out by WALTER (1956), yield does not even necessarily parallel CO_2 assimilation, for it depends on the manner in which the assimilate is used. In agreement with this, SIMONIS (1947), found that plants grown under conditions of drought assimilated more per unit of surface than those grown under conditions of optimum moisture, yet the latter yielded 50% more. This was due to use of the extra assimilate under conditions of drought to produce a larger root system. This resulted in a smaller assimilatory surface and therefore a smaller total yield per plant. In the same way Walter showed that the dry matter per unit surface is three times as great in drought resistant sclerophylls as in soft-leaved deciduous plants. Due to this use of larger quantities of assimilate to produce the leaves and due to the smaller leaf fraction of the total dry matter, the drought resistant sclerophylls therefore produce less substance and grow more slowly.

Walter (1955) points out that "poikilohydric plants" (those that take on the relative humidity of their environment) are not at all injured by complete drying to the air dry condition, yet they differ markedly in xerophily. Some cannot grow below a relative humidity of 99% (e. g. some bacteria), others can grow down to relative humidities of 88% (some molds). In fact, he actually classifies plants according to the difference between the osmotic potential that is optimum for growth and the maximum that they can tolerate before injury occurs due to dehydration. Those plants showing a small difference between the two he calls "stenohydric"; those showing a large difference he calls "euryhydric." *Therefore, resistance to growth stoppage by drought may be completely independent of resistance to injury by drought.* This fully disposes of Stocker's belief that incipient injury must accompany growth stoppage, and that further dehydration would have to extend the injury. On the contrary, growth stoppage may actually be a sign of resistance, e. g. when the plant enters its rest period. And the factors involved in the two kinds of resistance may be completely different. For instance, since growth stops when turgor pressure is zero, the ability to continue growth under adverse conditions depends partly on the plant's ability to maintain a high turgor pressure, e. g.. by accumulating large quantities of solutes as in some fungi, or by preventing water loss, as in some succulents. The ability to survive a low relative humidity may be completely unrelated to these mechanisms, e. g. in the case of plants that can be air dried without injury.

It is obvious, then, that the injurious effects of environment on growth and life must be separated if either or both are to be understood. The term environment resistance will, therefore be used specifically for the ability of a plant to resist protoplasmic damage when in an unfavorable environment. There is, in fact, ample precedent for confining the use of the term to actual damage. Other terms have long been in use to indicate the more general property of the plant to live and grow under adverse conditions, e. g., xerophily, thermophily, etc. When used in this exact sense, environment resistance cannot be measured by yield nor by the amount assimilated per unit dry matter as Stocker (1956) attemps to do. Both of these quantities can be lowered in many ways in the complete absence of protoplasmic damage.

Once environment resistance has been so defined, there are only two basic lines of defense open to the plant. This can be most graphically indicated by a teleological (and therefore metaphorical) question: Does the plant resist the unfavorable environment by erecting a barrier and keeping it out, or by preparing itself for the worst and letting it in? These two lines of defense may be called environment avoidance and environment tolerance, respectively. The latter is commonly known among practical men as hardiness (LEVITT 1956). Plants showing environment avoidance may conceivably be completely tender, i. e. they may possess no hardiness or tolerance whatsoever. Obviously, then, environment resistance may or may not involve hardiness. In fact, avoidance and tolerance may sometimes be mutually exclusive, e. g., if drought tolerance can be developed only as a result of moderate wilting, which would be prevented by drought avoidance. In any case, it is hopeless to look for a mechanism in any specific plant unless it is first determined which line of defense has been adopted.

The following are the main kinds of environment resistance:

1. Low temperature or cold resistance (including frost resistance),
2. High temperature or heat resistance,
3. Drought resistance,
4. Salt resistance,
5. Radiation resistance.

The first three are the only ones that have been intensively investigated. Terminology has been discussed in previous publications, but mainly with respect to drought resistance (MAXIMOV 1929, ARVIDSSON 1951, LEVITT 1956, PARKER 1956, STOCKER 1956). Most of the definitions are based on field experience, and strive for practical usefulness (VAN BAVEL 1953). They are, therefore, incapable of leading to a sound, quantitative treatment. Thus a definition of drought as a "period during which the soil contains little or no water available to plants" (PARKER 1956) is quantitatively meaningless since time is only a secondary factor and cannot of itself be used to measure drought. The actual drying power of the environment must be measured. Examples of the danger of such a time measurement can be found in much of the literature. OCHI (1952), for instance, exposed mosses to atmospheres varying from 53–77% relative humidity on one day, from 73–86% on another. Though a constant exposure time was used, the results are obviously not comparable. Similarly VAN BAVEL (1953) proposes that the incidence of drought be characterized by the number of days during its growing season on which soil moisture tension exceeds a value known to impede appreciably crop growth. Again, no exact measure of the degree of drought is used.

In order to obtain a quantitative measurement of environment resistance, it is necessary to consider constant, controlled conditions, such as can be obtained only in the laboratory, rather than the highly variable field conditions. Once such a quantitative system is developed, it may then be possible to extend it to cover natural conditions. But in the final analysis it will probably always be necessary to fall back on the more exact laboratory determinations. The following system is proposed.

2. Quantitative Aspects of Environment Resistance

Resistance is a quantitative character that may increase or decrease within a wide range. But it cannot be expressed quantitatively unless the potential causes of injury can be measured. These must therefore first be defined. On the basis of such definitions, the resistance can also be expressed quantitatively.

The frost, drought, and heat of a plant's environment, as well as the plant's resistance to these can be defined as follows:

1. Frost is a measure of the environment's freezing potential, i. e., its ability to induce ice formation. Since plants vary so much, pure water must be used as a reference point. Expressed by symbols:

$$F \propto T_f$$

Where F = frost, T_f = the freezing temperature in degrees below 0^0 C. (i. e. expressed as a positive value).

For most plant tissues, the time factor has little effect on the intensity of the frost, as long as the minimum temperature of the environment is kept constant for a few hours. For bulky organs, such as tree trunks, perhaps 24 hrs. would be required for temperature equilibrium to be reached (Reynolds 1939). For organs with large specific surface, such as leaves and twigs, one to two hours may be enough. In the case of subterranean plant parts, much longer is needed for equilibrium to be reached; but again, 24 hrs. is certainly enough for a potted plant.

Frost resistance can, therefore, be defined as the number of degrees of frost that can be survived by the plant. This is not quite synonymous with low temperature resistance, since the latter also includes chilling resistance, i. e., the ability to survive low, but non-freezing temperatures without injury.

2. Drought is a measure of the environment's drying potential, i. e. its ability to induce a net removal of water (with a reduction in the plant's water content). But this is more complex than in the case of frost, for the reduction in the plant's water content is the difference between the water removed by the air and the water supplied by the soil.

Since water removal from the plant is simply a diffusion process, it is proportional to the diffusion gradient. Again, using pure water as a reference point:

$$D_a \propto p_0 - p$$

Where D_a = atmospheric drought,
p_0 = vapor pressure of pure water,
p = vapor pressure of air.

Atmospheric drought is therefore proportional to the vapor pressure deficit of the air. Since relative humidity is more easily measured, it is more convenient to define atmospheric drought as proportional to the relative humidity deficit of the air, provided that a standard temperature is used.

But does this relation give a true measure of the drying potential of the environment? In the case of frost, it is easy to decide what should be chosen as a measure of the freezing potential of the environment, at least under artificial conditions when the plant is not exposed to light. Under natural conditions, however, the parts exposed to the sun may have much higher temperatures than those in the shade (LEVITT 1956). This complicating factor is even more important in the case of drought.

Since loss of water from the plant is a diffusion process, it is logical to suggest the vapor pressure gradient between the plant and its environment as a measure of the drying potential of the environment. And since the intercellular spaces of turgid leaves are essentially saturated with water, this gradient might be taken as the vapor pressure deficit of the surrounding air. But the energy for evaporation comes from the sun, which may heat the leaf above the temperature of its environment, raising its vapor pressure above that of the surrounding air at saturation. Therefore, under natural conditions, the vapor pressure deficit of the air is of little value as an indicator of water loss (THORNTHWAITE 1956); for the drying potential of the environment will then be much greater than its vapor pressure deficit.

But what actually happens to the leaf under conditions of strong insolation? If these conditions are constant for a short time, the leaf reaches a steady state. A balance is struck between the heat absorption from the sun and the heat release due to evaporation of water (and reradiation of heat). If the former is very high, this steady state will be at a temperature somewhat above that of the surrounding air. But if the leaf is able to transpire very rapidly, this temperature rise may be slight or non-existent. *Paradoxically, the leaf that loses water the most rapidly is the leaf with the smallest vapor pressure gradient;* for it is drawing on the soil water reserves and in this way keeps its temperature and therefore its vapor pressure down.

This leads to a logical method of evaluating the drying potential of the atmosphere. Let us take as the reference point the hypothetical leaf (or any other aqueous system) whose steady state temperature during insolation is equal to that of the surrounding air. In other words, its saturation vapor pressure is also equal to that of the surrounding air. For such a system the above equation holds, and the drying potential of the air is equal to its vapor pressure deficit. Any plant whose vapor pressure rises above that of the surrounding air at saturation will then have negative drought resistance, which may, of course be counterbalanced by other factors (see below). Consequently, though THORNTHWAITE is correct in saying that we cannot measure the drying potential of the air for any one plant by determining the vapor pressure deficit, nevertheless we *can* measure *drought resistance* on this basis.

From this point of view, the high transpiration rate of drought resistant plants that at first shocked botanists into disbelief, now becomes easily understood. It is actually essential, in order to maintain full leaf turgor. There is apparently no halfway measure open to the plant for maintaining leaf turgor under conditions of drought. Either it must lose practically no water, or it must lose tremendous quantities. Both methods work; but the former is more successful when the soil water reserve is very low, the latter when it is high enough, at least at depths that the roots reach. A plant that transpires at an intermediate rate cannot maintain full turgor, for the leaf temperature (and therefore its vapour pressure at saturation) will soon rise to such a point that the *net* transpirational loss will be large enough to produce wilting.

What method, then, should be adopted for measuring drought resistance? It is usually not possible under natural conditions to maintain a constant gradient for a long enough period of time. So the determination must be made under artificial conditions. It would, then, even be possible to maintain the theoretical vapor pressure gradient by allowing the plant to transpire adiabatically. This might be done by surrounding the plant with walls consisting of good radiators, thermostatically maintained at the same temperature as the air stream that is allowed to circulate over the plant. But this method would eliminate some of the very factors in drought resistance, e. g. the ability of a plant to maintain a low saturation vapor pressure by transpiring rapidly. Consequently, it would be far better not to maintain the leaf at the same temperature as that of the surrounding air, but to expose it to a constant artificial sun, i. e. to illuminate it with a strong light.

There still remains the difficulty of accounting for the role of the root system in drought resistance. It is, of course, impossible to duplicate field root conditions in a pot. It is, therefore, better to remove the roots altogether. In this way, it may even be possible to obtain a better idea of the range of drought resistance of a plant, than could be obtained by growing it in a single soil type. The method suggested is as follows:

(1) Cut the plant off at soil level.

(2) Vaseline the cut surface and stand the shoot in a drought chamber at a standard temperature, under a standard (strong) light source, for a standard time (e. g. 6–12 hrs. depending on the temperature) and repeat with similar groups of plants at a number of known relative humidities. Determine the relative humidity deficit that induces 50% killing. This is found by standing the droughted shoot in water (after cutting a fresh stem surface under water), covering the shoot with a bell jar (allowing air circulation from below) and leaving for about a week before estimating injury.

(3) Cut the surface of another shoot under water, place the shoot in the drought chamber, leading it through a hole in the base into a container of water. Repeat the above determinations.

In this way it should be possible to determine (a) the drought resistance of the plant below the wilting coefficient of the soil, and (b) the drought resistance of the plant at field capacity. These two values should give the minimum and maximum drought resistance with varying root and soil conditions.

Since the commonly used drought chambers never involve the above standardizations, it is no wonder that the results frequently fail to agree with field experience (LEVITT 1956). Even the above method may eliminate some of the factors in drought resistance (e. g. the ability of the plant to form new roots if the old have been killed by the drought). But it includes such a large fraction of them that it should give a good measure of field drought resistance.

3. Since the term heat resistance is really used synonymously with high temperature resistance, it is the maximum environmental temperature (in ^{0}C.) that a plant can survive for a standard time. The time for the attainment of equilibrium may be even less than in the case of frost, since there is no heat of fusion to slow up the process. But unlike frost resistance, once equilibrium has been reached, the time factor is always of decisive importance. A complication may enter if the plant is simultaneously exposed to drought. Consequently, heat resistance should be determined at water saturation, i. e. by plunging in water at the required temperature or by exposing to a saturated atmosphere at this temperature (see LEVITT 1956).

In all cases, resistance is best determined for the 50% point, i. e. the degree of frost, drought, or heat causing 50% killing. In some recent investigations, particularly with lower plants, attempts have been made to estimate injury from the change in respiration rate, e. g. in the case of drought resistance (RIED 1953) and heat resistance (see LEVITT 1956). Such results must be interpreted with caution. EAKS and MACHLIS (1956) have shown that cucumber fruit respire at a steadily decreasing rate at normal temperatures (13 to 30^0 C.); but at temperatures that result in chilling injury (0 to 5^0 C.), there is a sharp rise in respiration accompanying the injury. Not until general death of the tissue does the respiration decline.

With the terms defined, resistance to each of these potential sources of injury can be measured rather simply, and all gradations between minimum and maximum resistance can be found. Theoretically, frost resistance may vary from zero to 273^0 C. (i. e. exposure to absolute zero). Drought resistance may vary from a relative humidity deficit of 0 to 100% (i. e. exposure to an atmosphere over concentrated H_2SO_4). Heat resistance may vary from some value above 0^0 C. (perhaps 15—20^0 C.) to more than 100^0 C.

What, then, are the theoretically possible sources of such resistance?

3. The Nature of the Defense Mechanisms

In the earliest work, an attempt was sometimes made to find a single factor capable of explaining resistance in all plants. Nowadays, the pendulum seems to have swung to the other extreme. Some workers (PARKER 1956, STOCKER 1956) assume that since several factors have been found in different species, all must be present in any one species. There is, of course, no a priori reason for assuming this, and one species may conceivably survive because of a single resistance factor, another species because of a different one, still others because of a combination of factors. Whether any one organism can possess all the resistance factors is questionable. In the case of drought resistance, at least, some are mutually exclusive. A plant that survives due to avoidance may not be able to develop tolerance, because the very development of the tolerance in many cases depends on the actual exposure of the protoplasm to the environmental factor. Field determinations have born this out, for those plants whose saturation deficit during drought is slight, as a rule possess low critical saturation deficits (compare clover, alfalfa, and beet—Fig. 1).

Fig. 1. Comparison between the highest average natural water deficit (broken lines) and the critical water deficit (solid lines) for the leaves and shoots of different species. Expressed in percent of maximum water content.
(From Arvidsson 1951.)

The kinds of resistance should, therefore, not be classified according to the individual factors, but according to the nature of the defense developed by the plant. Once this is known, the actual mechanism by means of which the plant achieves this kind of defense may then be investigated, as well as the factors involved.

The possible lines of defense available to the plant may be classified as follows:

I. Frost resistance

A. Avoidance

1. The plant may conceivably avoid the frost by resisting the temperature drop to that of its environment in the three following ways:

a. Insulation against heat loss.

b. Accumulation of the heat of fusion of water on ice formation in non-living parts of the plants (e. g. vessels).

c. Accumulation of the heat released by respiration.

2. The plant may conceivably avoid the frost without preventing the temperature drop, by resisting ice formation in its tissues in the following two ways:

a. Undercooling (also called subcooling or supercooling).

b. Lowering of its freezing point.

B. Tolerance

1. The plant may tolerate freezing, provided the ice crystals are confined to a size below microscopic (Luyet's "vitrified" state—see Levitt 1956).

2. The plant may survive varying amounts of ice if this is confined to the intercellular spaces.

3. The plant may survive microscopically visible ice formation inside the cells.

II. Drought resistance

A. Avoidance. This may be achieved only if the plant is able to maintain its internal vapor pressure well above that of its environment during drought.

1. By resistance to water loss.

a. Due to stomatal closure. Cuticular transpiration is always somewhat slower than stomatal transpiration, though the difference between the two varies markedly from species to species.

b. Due to reduced cuticular transpiration. A less permeable cuticle than commonly found is necessary to derive maximum benefit from stomatal closure.

c. Other leaf modfications, e. g. reduced surface.

2. By increased water uptake. Many ways of achieving this are open to the plant.

a. From the soil, i. e. by removing a larger fraction of the soil's water. This may theoretically be due to (1) reduction of soil vapor pressure to a lower level, (2) extension of the root system through a larger volume of soil, (3) a deep enough root system to reach the water table.

b. From the air by periodic water absorption through the leaves. This may conceivably occur during periods of dew, fog, or light rain.

3. By improved translocation of absorbed water. This would enable the rate of absorption to keep up with high rates of water loss.

B. Tolerance. This is possible only if the plant's protoplasm is able to survive a reduction in its vapor pressure.

III. Heat resistance

A. Avoidance

1. By analogy with frost resistance, the plant may conceivably avoid injury by maintaining its temperature below that of its environment. Unlike the frost problem, however, in this case the plant temperature is nearly always *more severe* than that of its atmospheric environment. Consequently, we must consider any mechanism a defense *if it lowers the plant's temperature with reference to what it would be in the absence of that factor,* even if the plant's temperature is still above that of its atmospheric environment. The following four methods must be considered:

a. Insulation against heat absorption.

b. Loss of heat due to vaporization of water. This value is about seven times as high as the heat of fusion and may therefore be expected to have considerable effect.

c. Reduction in heat released by respiration. At high temperatures this may be expected to reach significant values. Any reduction in respiratory rate might therefore be protective.

d. Reduced absorption of radiation. It is this absorption of radiation that is responsible for the rise in temperature of the plant part above that of its environment. This may be prevented or reduced,

(1) by increased reflection of the incident radiation.

(2) by increased transmission of the incident radiation.

B. Tolerance. This means the actual temperature the protoplasm can rise to without injury.

A diagrammatic representation of these possible defense mechanisms is given in Table 2. The question is to what extent and in which plants are these theoretical expectations fulfilled. And when they are, what plant factors are responsible for them. These points will be considered separately for each kind of resistance.

Since environment tolerance (hardiness) has been dealt with in detail in another publication (Levitt 1956), this treatment of resistance will concentrate on avoidance, and will bring in only newer data or concepts of tolerance that have appeared since preparing the above publication. This means that environment injury will also be largely ignored; for an understanding of environment avoidance generally does not require any knowledge of the nature of the injury that is avoided. An understanding of tolerance, on the other hand, is impossible without a clear concept of how the injury may occur.

There is one exception to this rule. If avoidance involves disturbance of another vital function, an indirect kind of injury may arise, the nature of which must be understood in order to explain the plant's adaptation. Thus, many plants could conceivably avoid drought injury by keeping their stomata permanently closed during a drought. But this would stop photosynthesis, leading to a completely different kind of injury due to insufficiency of carbohydrates or other photosynthesis-dependent sub-

Table 2. *Theoretically possible kinds of environment resistance.*

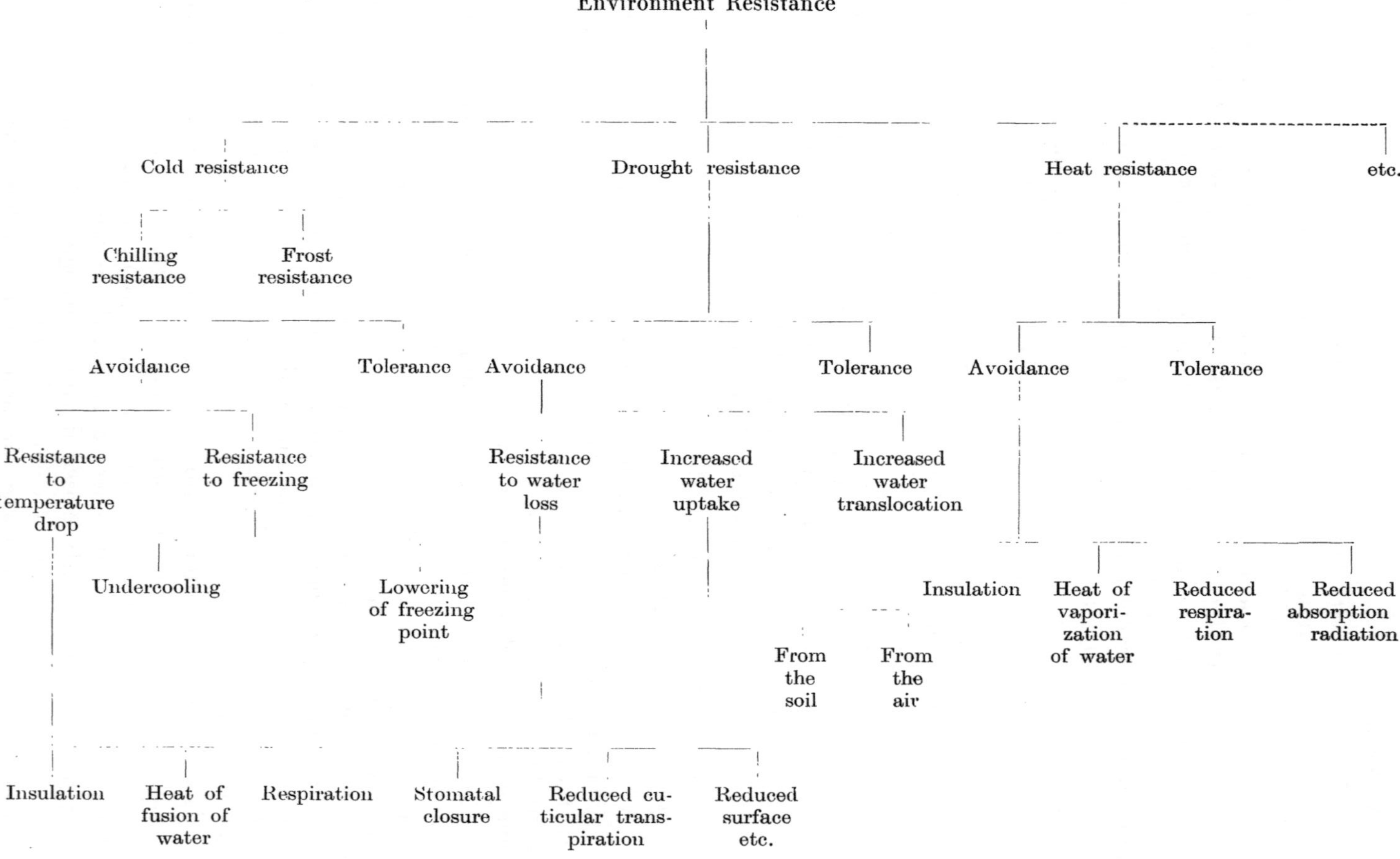

stances. Such indirect injury will have to be considered when dealing with environment avoidance, in order to determine whether any specific adaptation has been developed to counteract it.

4. Frost Resistance

In contrast to drought and heat resistance, there can be no ambiguity as to the meaning of frost resistance. No growth usually occurs at freezing temperatures, and the slight amount in exceptional cases cannot be a decisive factor in the adaption of a plant to its environment. Consequently, frost resistance can only mean the ability of the plant's protoplasm to survive freezing temperatures without injury. If only "winter hardiness" is measured, this may conceivably involve drought injury due to transpiration of the water without its replacement from the soil (Wilner 1955 b). But this cannot apply to relatively short time artificial measurements of frost resistance.

There is, however, one ambiguity in terminology. The term cold resistance or hardiness is frequently used synonymously with frost resistance or hardiness (Edgerton and Hoffman 1952, Chandler 1954, Edgerton 1954, Wood et al. 1950). This is in conformity with the German term "Kälteresistenz." Cold resistance should, however, include both chilling and frost resistance. In order to avoid confusion, the term should never be used except in this inclusive sense, or when it is not known which of the two more specific phenomena is involved. Consequently, the same criticism can be raised against its use by Kofranek (1952), Kondo (1952), and Andrew (1954), who were definitely investigating chilling resistance, i. e. resistance to temperatures above freezing.

A. Avoidance

There are two main methods by means of which resistant plants may conceivably avoid frost when exposed to freezing temperatures: 1. by resisting the temperature drop, 2. by resisting ice formation.

1. Resistance to temperature drop could conceivably occur in three ways:

a) Insulation. Though this is hardly possible in the case of leaves, it could be a factor in more bulky organs. Measurements do, in fact, show a lag in the temperature drop of tree trunks behind that of the air (Levitt 1956). But the difference is small and temporary.

b) Heat of fusion of water. This is also a factor in tree trunks. Just below 0° C., the temperature drop in a trunk temporarily comes to a stop even though the air temperature continues to drop (Levitt 1956). But this can be of value for only a short time and only near the freezing point. With a further drop in temperature, less and less water remains to be frozen; and if the protoplasm is dehydrated, it becomes a question of tolerance. Furthermore, this kind of resistance could be increased only by an increase in the plant's water content. Yet resistance has long been known to be correlated with a *reduced* water content (Levitt 1956). Heat of fusion of water therefore cannot be a significant factor in frost resistance.

c) Heat of respiration. Though respiration does occur even at subfreezing temperatures (LEVITT 1956), the rate is so slow that the heat evolved is negligible. This factor can therefore have no significance in frost resistance.

In actual fact, the temperature of plant parts may drop below the air temperature at times of greatest danger. JENNY (1953), for instance, showed that the temperature of apple and pear buds may rise above air temperature by as much as 4.7° C. when exposed to full sunlight; but at night, when air temperatures are lowest, it may drop as much as 2.5° C. below that of the air. SHAW (1954) found tomato leaf temperatures at night to be 3 to 4.5° C. cooler than that of the air when the sky was clear.

Fig. 2. Ice formation in veins of cabbage before occurrence in other tissues. (From ASAHINA 1956.)

From both theoretical considerations and experimental results, it is therefore evident that resistance to temperature drop is not a factor of importance in frost avoidance.

2. Resistance to ice formation

a) Undercooling (or subcooling) has long been known in plants (LEVITT 1956). LUCAS (1954) has recently reinvestigated the phenomenon in lemons. He found that the detached fruit vesicles were able to subcool to — 12° C., many of them freezing when the temperature dropped below — 18° C. Freezing could be induced by inoculating the cut stalk end with ice. No inoculation was possible at the sides. The freezing seemed to progress *via* the xylem vessels. In agreement with this conclusion, ASAHINA (1956) observed that on cold mornings ice formation in various crop plants was often confined to the veins (Fig. 2). This is perhaps one reason why undercooling is so rare in nature yet so easy to obtain with small sections devoid of large vessels. It may also mean that conifers with their narrow tracheids are more capable of undercooling than are angiosperms with their wider vessels.

LUCAS found the rate of spread of freezing in the lemon tree to be very rapid—in excess of 15 cm./min. at a subcooling temperature of — 5.2° C. This rapid spread led him to conclude that ice nucleation occurs at a single point in a subcooled tree. Though crystallization cannot spread across the membrane surrounding a vesicle, nor across the plasma membrane, LUCAS suggested that in subcooled lemons ice nucleation starts in the atmosphere near the fruit surface, then spreads through the peel to the whole fruit interior. His evidence for this was the following facts (Table 3):

a) The humidity of the surrounding air, or moisture on the fruit had no effect on the freezing point.

b) Fruit submerged in alcohol subcooled more than in air.

c) Fruit peeled before submersion in alcohol subcooled even more.

d) The prediction that rubbing a wet surface may greatly hinder subcooling was verified.

Table 3. *Effect of surface conditions on subcooling of detached lemons and lemon parts.*

(From LUCAS 1954.)

Group	Treatment	Ice Nucleation Point
		°C.
1	Freezing point of fresh fruits	—1.4 ± .05
2	Detached vesicles	below —12
3	Fruits in 90% relative humidity	—4.4 ± .20
4	Fruits in 90% relative humidity and stems heated	—4.2 ± .30
5	Fruits in 55% relative humidity	—4.0 ± .25
6	Fruits in 8% relative humidity	—3.9 ± .25
7	Fruits in 8% relative humidity and stems heated	—3.9 ± .29
8	Condensation on fruits	—4.3 ± .15
9	Fruits in alcohol	—6.1 ± .34
10	Half-peeled fruits in alcohol	—8.7 ± .40
11	Fully peeled fruits in alcohol	—8.5 ± .36
12	Fruits rubbed on water droplets	—2.6 ± .07
13	Fruits rubbed but not on droplets	—4.1 ± .18

This hypothesis hinges on the possibility of the ice nucleus being small enough to penetrate the relatively impervious lemon surface. LUCAS admits, however, that the surface of some herbaceous plants can prevent ice inoculation, and ASAHINA (1956) has found that intact flowers such as the Himalayan primrose and the late blooming dandelion freeze with difficulty even when dewdrops have frozen on their petals and their temperature is several degrees below their freezing point. It is, therefore, difficult to understand how the waxy lemon surface can fail to prevent ice inoculation of the interior. But other plant surfaces definitely do fail to act as a barrier. ASAHINA (1954) noted that subcooled potato sprouts are easily induced to freeze by frost formation on the leaf surface, in spite of the fact that drops of water the size of a plant cell do not normally freeze until a temperature of —15° C. is reached (ASAHINA 1956).

Whether or not such subcooling is a factor of importance in frost resistance is still uncertain. CHANDLER (1954) concludes that it is not, since it rarely lasts through a cold night. Presumably, most plants are like the above mentioned potato sprouts. It is still conceivable, however, that some exceptional plants may have the capacity to undercool to an unusual degree (see LEVITT 1956).

b) Lowering of the freezing point. This may possibly be a factor when very light frosts occur. A plant may avoid freezing at a temperature as low as —4° C. if it has an osmotic potential of 48 atms. Since it may be expected to undercool at least one or two degrees, the avoidance may extend still lower. However, most plants have freezing points closer to —1 to —2° C. In any case, the protection attained in this way is slight and of limited importance.

It cannot be denied, then, that frost avoidance may save plants from injury in certain specific cases, particularly where light frosts are involved. But in the great majority of cases that have been investigated, frost avoidance is too small to account for the plant's resistance. Consequently, most investigators have dealt with the major factor in frost resistance—tolerance or hardiness.

B. Tolerance (hardiness)

For about a decade, little work was published on frost hardiness (see LEVITT 1951). That the problem is still as important as ever is apparent from the resurgence of interest and the many papers now appearing. Even in the case of algal species prevalent on moors, it now appears that frost hardiness is one of the most important selective factors (HÖFLER 1951). One reason for the ten lean years was the realization that no new information could be expected without the development of new methods, or new lines of attack. The old methods and views had been repeated so many times that a stalemate had been reached. Unfortunately, the new generation of investigators seems unaware of this. As a result, many of the recent papers consist essentially of repetitions of the old lines of investigation, yielding the same results as were repeatedly obtained years before. These will, therefore, be reviewed very briefly.

The actual freezing temperature may not be the only factor that determines the injury. Killing may occur at higher temperatures if the plants are covered by an ice sheet than if frozen in air. In the case of alfalfa, injury was manifest only after 7–10 days; it could not be ascribed to any direct mechanical effect of the ice (SPRAGUE and GRABER 1940). After 20 days under the ice sheet, mortality was high. SPRAGUE and GRABER (1940) ascribed this to the accumulation of toxic products of aerobic and anaerobic respiration, since ice inhibits the diffusion of CO_2 and other respiratory products.

In favor of this explanation, plants immersed in water at 1° C. were injured at approximately the same rate and to the same degree as those frozen in ice at —4° C. When CO_2 was bubbled through the water, survival was reduced more rapidly and injuries were more intense. However, CO_2 was only one of several gases that appeared toxic. Later tests (SPRAGUE and GRABER 1943) showed that even at the highest concentrations of CO_2 used, a longer storage time was needed to cause injury than when the plants were frozen in blocks of ice. This led them to conclude that the external concentration of CO_2 was not directly toxic, but that respiratory compounds accumulated internally until they reached a toxic concentration. Thus injury similar to that produced by storage in blocks of ice could be induced by removal of both CO_2 and O_2.

Not all plants show the same sensitivity to encasement in ice. When frozen in this way at a temperature that is non-injurious when frozen in air (—3° C.), ladino clover died within 12–14 days, white clover survived four weeks or more (SMITH 1949). A relation to CO_2 concentration was again indicated, since this was consistently higher for the ladino clover. In general, SMITH (1952) found that the survival of ice encasement among legumes was approximately in the same order as their winter hardiness. He suggests that this may be due to a higher level of metabolic activity in the less hardy legumes. This agrees with BULA and SMITH (1954), who found that the rate of loss of carbohydrates in legumes during winter dormancy was inversely proportional to hardiness.

It has long been known that other factors besides ice encasement may also affect the degree of frost injury, e. g. the rate of freezing and thawing, and the length of time frozen (see LEVITT 1956). It is now apparent (LEVITT 1957 a) that even the post-thawing treatment may affect the degree of injury. There are thus four "moments of injury"—(1) during freezing, (2) after freezing equilibrium has been reached, (3) during thawing, and (4) after thawing.

The process of ice formation in a large number of plant species, both hardy and non-hardy, has recently been described in detail by ASAHINA (1956), and documented by an admirable and very extensive series of photomicrographs. His very careful and thorough observations completely confirm and extend those of previous investigators (see LEVITT 1956). Besides the commonly found types of intra- and extracellular freezing, he rediscovered a rarely observed ice formation between the cell wall and the unfrozen protoplasm. This was seen in the staminal hairs of *Tradescantia* as well as in several other plants. When the ice layer formed in this way, water was withdrawn from the cell interior and a "frost plasmolysis" frequently resulted without harmful effects. This is, of course, quite different from the early observations of "frost plasmolysis" in dead and thawed cells after extracellular freezing (see LEVITT 1956).

Any conditions that favored subcooling of the cell contents tended to lead to intracellular freezing. Two kinds were observed: (1) A rapid "flash," the cell contents suddenly becoming opaque due to an almost instantaneous formation of a large number of ice crystals throughout the vacuole and protoplasm; this occurred after pronounced subcooling. (2) A slower "non-flash" type due to a slower growth of larger crystals when ice formation occurred near the freezing point (i. e. with slight subcooling). Other differences occurred, but in nearly all cases intracellular freezing involved freezing of the protoplasm and death. The flash type resulted in freezing of the vacuole sometimes without freezing of the cytoplasm or nucleus, which looked unchanged even on thawing, yet were dead. Some unhardy cells froze intracellularly even if cooled slowly. But even these froze only extracellularly if inoculated with ice just below their freezing points. This is more likely to happen with large volumes of tissues than with sections, due to the large heat release on initial ice formation and the much slower rate of cooling in nature, both of which maintain the tissue near its freezing point. The non-hardy cells were injured even by the extracellular freezing.

Sometimes ice formation was prevented at the tonoplast, intracellular freezing occurring in the cytoplasm but not in the vacuole. This was often observed in very hardy cells. The ice was found to grow more slowly in the cytoplasm than in the vacuole. Barriers to ice formation occurred not only at the ectoplast and tonoplast, but also at the nuclear surface.

An even more rapid rate of freezing was used by Modlibowska and Rogers (1955), since they cooled the sections with the aid of dry ice. By means of motion pictures, the stages of freezing could nevertheless be studied. The freezing of mosses yielded an interesting variation from the process in higher plants. Intercellular freezing in the ordinary sense was impossible due to the absence of intercellular spaces. Yet ice was observed to form first in the cell walls, sometimes resulting in longitudinal cracks; presumably due to ice formation in the middle lamella separating the cells. If a cavity was produced in this way, it disappeared on thawing due to the reimbibition of the water by the cells.

The importance of speed of freezing and thawing has been reemphasized recently in another way. Luyet earlier showed that if the speed of both freezing and thawing are great enough, any kind of protoplasm can survive the lowest temperatures without injury, due to what he called vitrification (see Levitt 1956). Sakai (1956 c) now asserts that frost hardy plant tissues may be easily "vitrified" if subjected to gentle extracellular freezing. Almost all of the easily freezable water was in this way withdrawn from the cells of willow, poplar, and mulberry. When in this state, they could be plunged into liquid N_2 and kept there for as long as 60 days without injury. In the case of parenchyma cells of mulberry cortex, the gentle extracellular freezing had to be continued down to -30^0 C. before transferring to liquid N_2. But survival depended on rapid thawing—in warm (30^0 C.) water. If warmed slowly in cold (5^0 C.) water, the cells were completely killed.

The only possible explanation of these results is that the small amount of water left in the protoplasm at -30^0 C. is "vitrified" in the liquid air and crystallizes at temperatures between 0^0 and -30^0 C. if the temperature rises slowly enough. If, on the other hand, the temperature rise is fast enough, the "vitrified" water liquefies before it can crystallize. This would seem to indicate that even traces of ice crystals in the protoplasm can be fatal, though perhaps this is true only in the case of protoplasm already subjected to the strain of cell collapse due to extracellular ice formation.

There are still many apparent contradictions as to the effects of speed of freezing and thawing. On the one hand, Mazur's (1956) results with spores of *Aspergillus flavus* agree with the above results. When suspended in distilled water or in any of a number of solutions, and cooled rapidly to — 70° C., 75% survived rapid warming at 700° C./min. Only 7% survived slow warming at 0.7 to 0.9° C./min. The log of percent survival was a linear function of the log of rates of warming from 0.12 to 1000° C./min. Death during slow warming occurred primarily between —27° or —20° C. and 0° C. It did not occur during thawing of ice around the spores. Injury was not due to the longer exposure to low temperature during slow thawing, for if exposed to a constant temperature between — 70° and 0° C. for as long as 2 hrs., they were less injured than spores warmed slowly. On the other hand, later results by Mazur et al. (1957) seem to contradict these results, for when *Pasteurella Tulariensis* was cooled slowly to — 15 to — 75° C. in sugar solutions, slow warming resulted in more than 60% recovery in all cases. Rapid cooling decreased recovery from 60% at — 15° C. to 40% at — 75° C. But in this case, rapid cooling meant 16–50°/min., slow cooling (and warming) meant 1°/min. It is obvious that some standardization of terminology is needed. The following may suffice, for rates of cooling, warming, freezing or thawing, at least for artificial freezing tests using small samples.

Slow rates: 1—2° C. or less/min.

Rapid rates: 5—20° C./min.

Ultrarapid rates: 100° C. or more/min.

Intermediate rates would have to be referred to as such. It would then be safe to generalize, for instance, that rapid freezing causes more injury than slow freezing, and that ultrarapid freezing prevents injury.

But even this terminology would not hold for whole plants frozen naturally, since the temperature drop in nature is much slower than the 2° C./min. used in these experiments. Biel et al. (1955) inserted thermocouples into the stolons of ladino clover and found maximum cooling rates of 10° F./hr. Even these rates were fast enough to produce rapid cooling injury. Sprague (1955), observed more injury in ladino clover and alfalfa when cooled from 36° to 10° or 5° F. at a rate of 7° to 10° F./hr. This rate is incapable of producing rapid freezing injury in the case of sections (Levitt 1957 a) since it is slow enough to permit the small amount of ice formation to occur outside the cells. The large amount of ice that must form in the whole plant requires a much longer time and therefore cannot be completed before a further temperature drop. In other words, a rate of freezing that is rapid enough to cause intracellular freezing in the case of the whole plant is slow enough to permit only extracellular freezing in the case of sections or small amounts of living material.

The Japanese Institute of Low Temperature Science at Hokkaido University recently thoroughly reinvestigated the "double-freezing point" of Luyet and other workers (see Levitt 1956). Aoki (1948) confirmed the earlier findings of Luyet and his coworkers; the first freezing point could be obliterated by concentrating the surface solution in the damaged cells of a tissue piece; it could be enhanced by diluting it. He, therefore, concluded that the first freezing point depends solely on the conditions of the surface layers of tissue. The second freezing point, on the other hand, becomes irregular only when the frequency of intracellular freezing ("cell flashing") changes irregulary or rapidly. From this he concluded that the first freezing point is due to the freezing of the damaged surface layer of tissue, and the second freezing point to the freezing of the inner tissues. Whether or not the two are separated, depends on the manner of freezing of the internal tissues. If they freeze easily, the freezing progresses continuously from the damaged surface layer throughout the living internal tissues and the second freezing point is obliterated. If the living tissue freezes with difficulty, it is subcooled following the freezing of the damaged surface layer and the second freezing point is distinct.

Aoki later (1950) attempted to find a relation between the shape of the freezing curve (temp. *vs.* time) and the mode of freezing of plant tissues. Tissue pieces ($5 \times 5 \times 1$ mm.) of ten common vegetables were subjected to freezing at a definite rate of cooling (2–3° C./min.). The tissue temperature was measured by means of a thermocouple. Three types of curves were obtained: Type I had two sharply defined freezing points and the freezing was mainly intracellular. Type II showed only the first freezing point and both intracellular and extracellular freezing occurred, though the former was not as continuous as in I. Type III showed two freezing points, but the second was not sharp; freezing was mainly extracellular. Types I and II occurred in completely unhardy tissue (melon and cucumber fruit), Type III in tissues that were at least capable of hardening (e. g. cabbage). This relation to hardiness was more fully investigated by Aoki et al. (1953). In the hardened state, the living cells of the petiole and root of red beet froze extracellularly and the two freezing points were not clearly distinguishable. In the dehardened state, intracellular freezing occurred suddenly after subcooling and there were two distinct freezing points. It is, therefore, sometimes possible to distinguish between hardy and non-hardy tissues by the shape of the freezing curve. Asahina (1956) also found hardy cells resistant to intracellular freezing. These results confirm Siminovitch and Scarth's earlier findings that hardy cells avoid intracellular freezing better than do non-hardy cells (see Levitt 1956).

Aoki's explanation of the double freezing point is based on his assumption that rinsing a piece of tissue with water or a sugar solution affects only the damaged surface layer of cells. But this assumption is invalid, for the turgor of the internal cells must also be affected. He would also have to explain why the ice that forms in the damaged surface cells does not spread throughout the tissues. According to Asahina (1956), Ulrich has explained the prevention of intracellular freezing in hardy cells by an increase in glycoproteins, though Levitt (1954) failed to find any such increase. An earlier explanation (see Levitt 1956) was the increased cell permeability to water which must permit a more rapid exosmosis of the intracellular water to the extracellular ice loci. But even when ice formation is solely extracellular, Aoki was still able to detect a double freezing point (Type III), though less distinct than when ice formed intracellularly (Type I). His hypothesis fails to explain this. The following turgor theory is therefore suggested.

Ice forms first at the cut surface of the piece of tissue. Since it is so small the whole piece of tissue is essentially in vapor pressure equilibrium; and since there is no barrier to the propagation of the ice throughout the intercellular spaces, the ice spreads forming a continuous thin layer on all the cell surfaces. If the tissue has been rinsed with water before freezing, it is in the turgid state at the instant of freezing and the freezing point is therefore *higher than the freezing point of the cell sap.* This results in a high first freezing point somewhere between 0° C. and the freezing point of the cell sap, depending on the balance between the temperature gradient and the amount of water able to freeze in the tissues. But this initial, relatively rapid, ice formation at the cell surfaces removes enough water from the living cells to cause loss of turgor. The result is a *marked* lowering of the freezing point due to the double effect of turgor loss and increase in cell sap concentration. Further ice formation is therefore much slower until all turgor is lost. At this point ice formation can occur much more rapidly. As an example, a fully turgid piece of tissue with a freezing point lowering of — 2° C. would have about 10% of its water frozen between 0° C. and — 2° C., 45% frozen between — 2° C. and — 4° C. This would give the appearence of two freezing points. The double freezing point could, therefore, occur whether freezing is intracellular or extracellular, though showing up more sharply in the former case due to greater subcooling. It should be pointed out that if AOKI's explanation is correct, the double freezing point is an artifact that occurs only in cut tissue. If the turgor theory is correct, it can occur also in normal, turgid, undamaged tissue, though perhaps depending on the rate of temperature drop.

All these observations by the Japanese workers fully confirm the earlier conclusions (see LEVITT 1956). (1) The plant does not tolerate intracellular ice formation (except at the extremely rapid rates of temperature lowering used by LUYET). (2) Hardy plants are far more capable of avoiding intracellular freezing than are non-hardy plants, presumably due to their higher cell permeability. (3) Hardy plants are more tolerant of extracellular freezing than are non-hardy plants; and the tenderest plants are killed by even the slightest amount of extracellular freezing. Thus only one kind of frost tolerance really exists under natural conditions, i. e. tolerance of extracellular ice formation.

Much of the recent work on hardiness has been practical in nature, with two major aims; (a) to determine the relative hardiness of species or varieties, and (b) to increase their hardiness artificially.

(a) The methods used to measure frost hardiness have remained essentially unchanged (e. g. AMIRSHAHI and PATTERSON 1956), except for the marked tendency, especially among horticulturists and agronomists, to adopt the conductivity method of measuring injury (CARRIER 1951, COOPER et al. 1954, EMMERT and HOWLETT 1953, WAY 1954, WILNER 1955 a, BULA et al. 1954, 1956, etc.). The author has already pointed out (LEVITT 1956) that these results would be much more useful if the frost killing point were determined. This could easily be done by determining in each case the conductivity from completely frost killed plants. The temperature resulting in half this conductivity could serve as the frost killing point. An attempt to determine injury by use of 2, 3, 5-triphenyltetrazolium chloride proved unsuccessful (CARRIER 1951). TORSELL and HELLSTRÖM (1955), however, used this dye in conjunction with indigo carmine to distinguish living cells from dead to estimate the injury caused by freezing. The plasmolysis resistance method has proved to give a perfect measure of frost hardiness in the case of mulberry (SAKAI 1955 b, 1956 b). So far, this method has never failed.

A new "rapid method" for determining winter hardiness in alfalfa has been described (RODGER et al. 1957). The seeds were germinated in solutions of NaCl or sucrose of known osmotic potential. As the osmotic potential increased, the rapidity and amount of germination decreased. This decrease was more marked in the seed of hardy varieties. Eight varieties were in this way arranged in the order of their hardiness. It is a little difficult to understand how such a method can possibly work since it is not based on any known hardiness factor and in fact determines the quantity before it has actually developed in the plant. Furthermore, it is in direct opposition to the repeated claims of European workers that increased frost resistance can be obtained by selecting seed that germinate in solutions of high osmotic strength (FRYXELL 1954).

HELLSTRÖM (1956) has attempted to evaluate hardiness on a three -component diagram, using non-reducing sugars, reducing sugars, and soluble N. Whether or not this gives a better measure than sugars alone, remains to be seen.

Morphological characters associated with hardiness are still being sought. In the case of alfalfa (SMITH 1955 a), varieties arranged in the order of the underground development of crowns were also in the order of their hardiness. Winter injury was also directly related to the height of the plants (SMITH 1955 b). SAKAI (1955 a) was able to estimate the frost hardiness of mulberry by the length of the corkless part of the twigs in early autumn, since it is formed as growth slows down.

Seasonal changes in hardiness have been followed for a number of plants using the well established artificial freezing tests. The hardiness of mulberry cortex increased from —5° C. in September to —35° C. in December, this maximum continuing through February (SAKAI 1955 b). The above ground parts may show as much as 11.5° greater hardiness than the below ground parts (TILL 1956). These differences increase in winter, decrease in summer. Even the above ground parts of these evergreen species may show differences of 5° C., the younger leaves being less resistant. The yearly amplitude among 24 herbaceous and 19 woody species varied from 9 to 36° C. for the above ground parts and 4.5 to 8.5° C. for below ground parts. The needles of *Pinus ponderosa* and *Abies grandis* showed a slow increase in hardiness during late summer and early autumn, a sharp increase during early November which coincided with several nights near or below freezing (PARKER 1955). During January they withstood at least —55° C.

Alfalfa varieties developed resistance faster and to a greater degree at the higher latitude of Alaska than in Wisconsin (BULA et al. 1956). In the case of the less hardy variety (Arizona common) the difference was slight and it was killed in both latitudes.

Several recent investigations have used artificial freezing tests to measure the relative hardiness (Fig. 3) of a number of varieties (EMMERT and HOWLETT 1953, COFFMAN 1955, RACHIE and SCHMID 1955, JOHANSSON et al. 1955, AMIRSHAHI and PATTERSON 1956). This method has even been used in the field, in the case of 2-year old grapefruit trees (COOPER et al. 1954), as well as for winter turnips and rape and for rye (JOHANSSON and TORSSELL 1956). In the latter case, temperatures as low as —28° C. were obtained over an area of a square meter. Greatest frost sensitivity was found at the flowering stage, e. g. rye was injured at 0 to —2° C. These artificial freezing tests, as in previous work (LEVITT 1956), have given good agreement with field experience. In the case of 53 apple varieties, however, this was true only when they were conducted in the fall (EMMERT and HOWLETT 1953). Varieties that were very hardy in the fall proved relatively more tender in winter, while the medium hardy and tender varieties of the fall became correspondingly more hardy relative to the others. In general, varietal differences diminished as the season progressed. Similar differences in the time of development of maximum hardiness have been found in legumes (BULA and SMITH 1954). Hardiness developed more rapidly in sweet clover than in alfalfa, and also reached a higher level. In red clover, it developed later and more slowly than in either of the other two and a lower maximum was reached. Hardiness began to decrease after mid-February in red clover and alfalfa, but was retained longer in sweetclover.

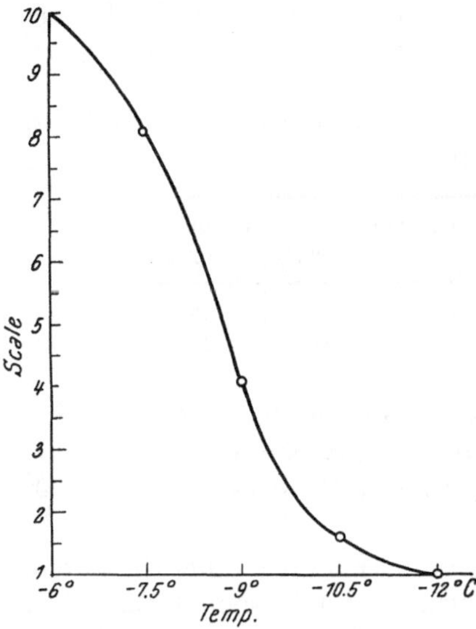

Fig. 3. Frost hardiness of plants frozen at five different temperatures. Ordinate is hardiness scale used by Swedish investigators. 10 = uninjured, 1 = completely killed. (From JOHANSSON et al. 1955.)

Stage of development has again proved important. Seedlings of *Lespedeza striata* in the two-leaf stage showed superior hardiness which was lost with development of the third and fourth leaves (TABOR 1951). Underdeveloped parts of the mulberry tree failed to develop frost resistance even if exposed to 0° C. for 10 days (SAKAI 1955 a). On the other hand, when development advanced to the point where growth activity decreased, some degree of frost resistance was developed even without exposure to low temperature. Varietal differences in hardiness can also sometimes be related to development. Thus, the upper twigs of non-hardy varieties of mulberry are in an underdeveloped stage in early autumn, but in hardy varieties they have already completed their development. Similarly, among seedlings of medium red clover, plants that flowered winter killed more than non-flowering plants (SMITH 1957).

Cut twigs of mulberry hardened artificially at any temperatures below 10° C., the effectiveness increasing with the time and with the temperature below 10° C. (SAKAI 1956 b). In agreement with the earlier consensus of opinions (see LEVITT 1956), alternating temperature (16 hrs. at 0° C., 8 hrs. at 15° C.) proved no better than continuous exposure to 0° C. At subzero temperatures, if the period was long enough and freezing was extra-cellular, the hardening process continued.

SAKAI (1956 a) has clearly shown the relation between length of time frozen and amount of injury. The time at —10° C. survived by mulberry

Table 4. *The time (days) survived by twigs frozen at different temperatures.* (From SAKAI 1956 a.)

Tree	Temperature frozen and stored at			
	−5° C.	−10° C.	−20° C.	−30° C.
Mulberry	180	30	10	½
Poplar	360	90	50	20

and poplar increased with increase in hardiness, reaching 30 days for mulberry, 80 for poplar. —5° C. was survived for fully 360 days by poplar. Browning due to injury occurred after 60, 30, and 10 days at —10, —20, and —30° C., respectively (Table 4).

The well known fact that dehardening occurs at high temperatures has received some attention (EDGERTON 1954, BRIERLEY and LANDON 1952, 1954). While a 4-day exposure to warm temperatures failed to induce much loss in hardiness, a similar 7-day period caused an appreciable loss. This was true even when such warm periods occurred naturally in the fall.

LARCHER (1954) investigated several Mediterranean evergreens. The difference between summer and winter frost hardiness was, in general, much less than in the case of more northern regions, though nevertheless distinct. The difference was usually greatest for most hardy species; but there were two pronounced exceptions. In contrast to all the other species, the two palms *Chamaerops humilis* and *Trachycarpus fortunei* were able to survive —10° and —13° C. respectively in summer and practically the same in winter. *Olea europaea,* on the other hand, required some exposure to cool temperature before becoming frost hardy. SATO et al. (1952) were able to increase the hardiness of hardwood and conifer seedlings during July and August by means of a 20 to 30 day exposure to thort days (10 hrs.). CHESSIN (1953), however, was unable to increase the hardiness of cheatgrass (*Bromus tectorum* L.) seedlings by exposure to short days. This is perhaps not surprising in view of the kind of plant and the short test period.

LARCHER (1954) adopts PISEK's distinction between "resistance stable" and "resistance labile" plants. In the former, hardiness is largely independent of external factors; in the latter, it depends on the weather of the moment. The difference was shown by artificial temperature changes. Cooling failed to differentiate between the two, since it was impossible in this way to increase the resistance more than 1° C., either at the time of lowest resistance in summer, or of highest resistance in winter. But they all reacted strongly to warm treatment in winter. Within 4 days at 20° C., 30–65% of the increased resistance was lost, the most stable species showing the least change. Even more striking was the comparison between the hardiness of greenhouse grown plants during winter and that of outdoor plants during summer. The difference varied between 0 and 50% (Table 5) of the normal increase in the open.

It may be that the difference between these "resistance stable" and "resistance labile" plants is not due to a purely autonomic or endogenous

Table 5. *Frost hardiness during winter of plants kept in a warm greenhouse.* (From Larcher 1954.)

Species	Freezing Temperature (°C.)		Increase in hardiness over plants in tho open during summer	
	survived without injury	15% killing	°C.	Percentage of increase in open
Olea europaea	−5.5	−6.5	0.0	0
Arbutus Unedo	−6	−7	1.0	20
Cupressus sempervirens	−9	−10	2-0	25
Viburnum Tinus	−6	−7	2.0	30
Laurus nobilis	−6	−9	2.5	50

change. Perhaps the former group have their hardiness controlled by photoperiod, the latter by temperature. Possibly the two palms that showed no difference in hardiness between winter and summer belong to a third group that are not affected by either of these factors.

The importance of protective solutions, long ago proved for higher plants, has now been extended, to microorganisms. Mazur et al. (1957) found that if *Pasteurella Tulariensis* is frozen in glucose or lactose solutions down to temperatures of —75° C., more than 60% recovered if the solutions were hypertonic (0.30 or 0.62 M) more than 18% if the solutions were hypotonic (0.15 M). In gelatine-saline, recovery was only 0.01%. Because the hypotonic solution also gave protection, they concluded that "It is apparent that if sugars reduced or prevented intracellular freezing, they did not do so by dehydrating the cells." They apparently failed to realize that as freezing progresses, the external hypotonic solution becomes more concentrated until it is hypertonic and dehydrates the cells. Åkerman (see Levitt 1956) long ago showed that plasmolysis does occur when cells of higher plants are frozen in hypotonic solutions. It is even possible to reduce the amount of injury by use of protective solutions immediately after thawing (Levitt 1957 a). This is a true protection that can be obtained with whole leaves as well as with sections.

One of the modern trends that cannot be found in the earlier work is the inevitable attempt to increase hardiness by means of sprays that have produced such magical results in other phases of physiology. EDGERTON and HOFFMAN (1952) thinned peaches with dinitro blossom sprays and the Na salt of naphthaleneacetic acid (Na NAA). The following winter, periodic freezing tests showed that the dinitro thinning at blossom time favored subsequent hardiness of the fruit buds more than did the post blossom (1 month after blossoming) Na NAA thinning, though the fruit reduction was comparable. But the latter was just as effective in increasing hardiness as was hand-thinning after the June drop. This would seem to indicate that the sprays did not have a specific effect, but merely operated by reducing the drain on reserves. MODLIBOWSKA and RUXTON (1954) sprayed raspberry plants at "bud burst" with 750 ppm. maleic hydrazide, and subjected them to low temperature at the "green bud" and "open flower" stages. This delayed the growth of many buds, particularly the more advanced ones, and killed others. There were three effects on hardiness: (1) the frost damage increased where growth inhibition was strong, (2) some degree of resistance was developed in those buds not too strongly inhibited, and (3) the delay in blossoming tended to reduce frost damage. The net effect, however, was a reduced yield. MORETTI (1953) was unable to delay bud development by any of the winter sprays he tried; therefore none protected against late frost.

In two cases, sprays have been asserted to increase hardiness. CORNS and SCHWERDTFEGER (1954) grew beet seedlings in vermiculite moistened with (a) water, (b) 4–8 ppm. Na salt of TCA, (c) 2,2-dichloropropionic acid. The shoots were then severed and exposed to —10° C. for 5–10 mins. A highly significant increase in survival was obtained in the case of the chemically treated. This, of course, yields no information whatever about frost hardiness, since the exposure was far too short to permit freezing equilibrium, if freezing occurred at all. It may possibly be a case of chilling injury, or perhaps frost avoidance, due to a lower freezing point. Further evidence of this conclusion was produced later (MILLER and CORNS 1957) by showing that the treatments effective in "improving cold resistance" actually decreased desiccation resistance and therefore must have also decreased frost hardiness. 40 days after full bloom, CRANE (1954) sprayed apricot trees with 100 ppm. 2,4,5-trichlorophenoxyacetic acid; 15 hours later there was a frost of 32° F. for 3 hrs., 31° F. for 1 hr. This treatment supposedly increased frost hardiness. Actually, it is obviously just a case of protection from freezing due to the heat of fusion of the water drops (see ROGERS et al. 1954), since no water sprayed controls were used. COOPER and PEYNADO (1955) applied plant regulators to leaves and roots of grapefruit trees in the field during early December. 2,4-D, 2,4,5-T, MH, Dalapon, and several other substances all had little or no effect on frost hardiness; but neither was growth affected except insofar as they produced injury. LONA et al. (1956) sprayed seedlings of *Brassica nigra* with aqueous solutions of different substances for 4–7 days then exposed them to —3 to —4° C. for 1 to 1½ hrs. More than 60 tests made with 4500 plants yielded the following consistent results. All substances that inhibited plant activity (tannic acid, gallic acid, malonic acid, eosine) increased frost resistance; growth stimulating substances (indole acetic acid, naphthalene acetic acid) reduced frost resistance. LONA (1956) further found that gibberellic acid decreases frost resistance.

The best results seem to have been obtained with trees. SAKAI (1957) sprayed 0.1 to 0.5% maleic hydrazide on mulberry leaves at the end of August. Growth was nearly completely stopped within 30 days, starch accumulated, and the osmotic potential of the parenchyma cells of the cortex increased over the controls. At the end of September the treated plants were more frost resistant than the controls. They were also able to increase in hardiness when exposed to 0° C. for 7 days—a treatment that failed to affect the controls at this time. KONOVALOV (1955) used extracts of winter hardy plants (*Bergenia crassifolia* leaves) to treat seeds and green plants (stock and cabbage), and reports increased frost resistance as a result of the treatment. Other changes, such as increased carbohydrate content were noted.

A completely different protective substance has been used by JEREMIAS (1956). On the basis of earlier, unpublished experiments by ULLRICH, he allowed wheat grains to swell in water for 48 hrs., then continued their germination in solutions of D-ribose. Controls were allowed to continue germination in isosmotic glucose solutions. A little thymol was included to prevent growth of microorganisms. After 15 days, 2 leaves had formed, and the plants were tested by transfer to a cold chamber. In all cases, the seedlings germinated in the ribose solution showed better survival (Table 6).

It is unfortunate that JEREMIAS used such short freezing tests. The maximum freezing time was 60 mins. and differences were found even after 20 mins. It is obvious that freezing equilibrium could not possibly be attained in this short time. Consequently, the differences may simply have been due to undercooling or to some other factor. It is doubtful whether they can be ascribed to hardiness differences on the basis of these tests.

It is quite apparent that little success has yet been achieved in attempts to increase frost hardiness by means of artificially applied chemical growth regulators and related substances.

The effects of fertilizers on frost hardiness are once again being investigated, and the results are similar to those in the older literature. LIVINGSTON and SWINBANK (1950) obtained increased injury in wheat with high fertility; WAY (1954) increased apple tree injury with N applications. GESSNER and ZWERENZ (1950) were similarly able to increase hardiness in a number of species by lack of N. Unfortunately, several of the species mentioned (e. g. *Tradescantia, Helianthus annuus*) are completely frost sensitive, so it is doubtful whether frost hardiness was involved. Increases in hardiness due to K fertilization have been recorded for hardwood and conifer seedlings (SATO et al. 1952) for one-year old tung trees (SHEAR 1953), and for pecan trees (SHARPE et al. 1954) at least under conditions of low K. Even Zn has been asserted to increase the hardiness of tung trees (SHEAR 1953).

Factors that have long been associated with hardiness are again receiving attention. A relation between sugar and hardiness has been found in sugar beets (Wood et al. 1950), Marsh grapefruit (Sharples and Burkhart 1954), legumes (Wood and Sprague 1952, Bula and Smith 1954, Ruelke and Smith 1956), wheat (Johansson et al. 1955), turnips and rape (Hellström 1954 a, b, 1955, Hellström and Torssell 1955), and, judging by the winter increase in osmotic potential, in fir (Kato 1951) *Saxifraga stolonifera* (Yae 1955), and mulberry (Sakai 1955 b).

The following are some of the details from the several above investigations. In agreement with earlier observations, spruce (in contrast to fir) had a high but constant osmotic potential from November to March. The starch to sugar conversion in winter occurred at a much higher temperature (55° F.) in grapefruit than is commonly found. There was a reduction in total available carbohydrates during winter in the legumes, but less so in the more hardy species, indicating a relation between low winter metabolic activity and hardiness. Hardiness of winter wheat actually decreased in darkness (Fig. 4), presumably because of the carbohydrate loss. (Johansson et al.); yet an increase did occur at first if the temperature was low. Sucrose was the dominant sugar in hardened plants, though high polymer sugars tended to increase in the late hardening stages. This agrees with the results of Gunar and Sileva (1954). They found that raffinose and a less mobile "tail" substance (determined chromatographically) developed in winter wheats during the second stage of hardening. Under natural conditions, these substances are present during winter, decrease to small amounts at the end of April, and disappear completely by the middle of May. Glucose, fructose, and sucrose accumulate earlier in the hardening period.

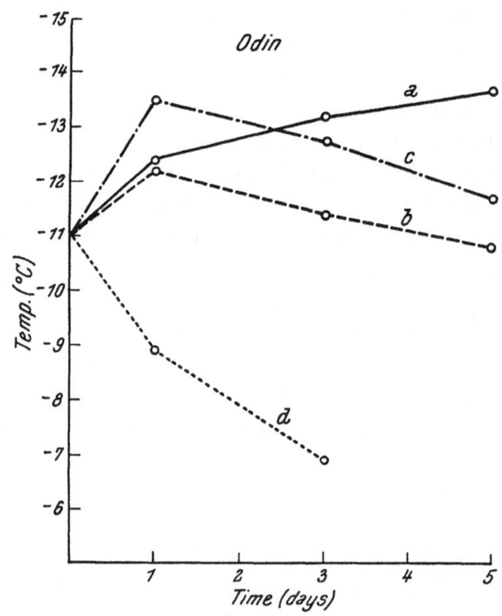

Fig. 4. Effect of darkness on hardiness of wheat. *a* +3° C. and light, *b* +3° C. and darkness, *c* —3° C. and darkness, *d* +20° C and darkness. (From Johansson et al. 1955.)

The accumulation of sugars does not always parallel hardiness. The least resistant alfalfa variety of the three tested by Bula et al. (1956) showed the highest level of total available carbohydrates both when grown in Alaska and in Wisconsin; the hardiest variety (*falcata*) accumulated the least. Larcher (1954), failed to detect any association between frost hardiness of Mediterranean evergreens and osmotic value. The slight change that did occur was, in fact, in the wrong direction. Presumably the freezing point lowering method was used, and the results may have been partly due to the reduction in water content.

Table 6. *Frost resistance of plants pretreated with ribose and glucose solutions respectively.*
(From Jeremias 1956.)

Variety and temperature	Solution	Number survived after		
		20 mins.	40 mins.	60 mins.
Hauter's II	Ribose	20	20	20
−3° C.	Glucose	19	19	17
Hauter's II	Ribose	19	18	17
−5° C.	Glucose	18	13	11
Hauter's II	Ribose	20	14	11
−8° C.	Glucose	18	10	3
Hauter's II	Ribose	19	14	6
−12° C.	Glucose	16	5	0
Criewener 192	Ribose	20	20	19
−8° C.	Glucose	20	18	12
Criewener 192	Ribose	20	19	14
−12° C.	Glucose	20	15	9
Taca	Ribose	20	19	14
−8° C.	Glucose	19	12	6
Taca	Ribose	20	16	11
−12° C.	Glucose	18	10	4

The role of sugars in hardiness has been explained on a purely osmotic basis (Levitt 1956). In an attempt to investigate this osmotic effect quantitatively (Levitt 1957 b), hardy cabbage plants were infiltrated with glycerine solutions. Frost hardiness was in this way increased to the extent

Table 7. *Effect of infiltration with M glycerine on frost killing temperature (T_k) of cabbage leaves.*
(From Levitt 1957 b.)

Time hardened (weeks)	Treatment	Osmotic Potential (atms.)	$\frac{C_{f2}}{C_{f1}}$	T_k (°C.)	
				determined	calculated
1½	control	20.25		−7	
	infiltrated twice	27.0	1.34	−11	−10
2½ (whole plant)	control	21.9		−7	
	infiltrated once	25.2	1.15	−9.5	−8.5
8	control	25.2		−12.5	
	infiltrated once	32.3	1.25	−15.5	−15.5

predicted by theory from a purely osmotic effect (Table 7). Earlier attempts to obtain even a qualitative agreement had failed because non-hardy plants were used and the expected effect in these plants is too small to be detected (see LEVITT 1956).

That carbohydrates may play some role in hardiness other than a purely osmotic one has long been suspected (see LEVITT 1956). TONZIG (1941) has suggested that mucoproteins, consisting of proteins carrying polysaccharides (mucilages, gums) as prosthetic groups play an important

Table 8. *Carbohydrate contents (%) of protein fractions of cabbage seedlings and potato tubers. Each value an average of 4–10 determinations.* (From LEVITT 1954.)

	Cabbage seedlings		Potato tubers	
	Unhardened	Hardened	10 days at 26° C.	3° C.
Mitochondria			8.3	10.2
Microsomes	17.5	17.1	6.8	9.8
Acid-insoluble proteins	5.7	6.1	5.0	6.0
Globulins	10.8	8.4	4.5	3.9
Albumins	28.8	22.3	6.7	6.0

role in the life of the plant. They are associated with and dispersed in the protoplasm but differ from the glycoproteins of animals characterised by the presence of aminosugars, glucosamine or chondrosamine. The major evidence that he produces for their existence is based on observations of fixed and stained cells. He suggests that these mucoproteins can dissociate into the protein and polysaccharide by the process of mucophanerosis. The two products are less and more hydrophilic, respectively, than the mucoprotein from which they are liberated. This results in a tremendous increase in swelling power of the polysaccharide. The consequent increase in hydrophily of the cell regulates the turgor, mechanical resistance, movements, etc. of the plant. It also explains the osmotic and viscosity changes that have been found associated with frost and drought resistance. LEVITT (1954) attempted to test this hypothesis by isolating the protein fractions of hardened and unhardened cabbage leaves and analyzing them for carbohydrates. No relation could be found between the carbohydrate content of the proteins and hardiness. It was, however, found that non-hardy potato tubers contained proteins with lower carbohydrate contents than those of hardy cabbage (Table 8). But this would seem to be the opposite of TONZIG's "mucophanerosis."

JEREMIAS (1956), on the other hand, has produced evidence in favor of the concept, at least in winter wheats. He found grana in the epidermal and mesophyll cells, after alcohol fixation, that were always more numerous in frost hardened plants. These grana were stainable with hematoxylin or toluidine blue, they were not destroyed by pepsin, ribonuclease, or hyaluronidase, but they disappeared on treatment with perchloric acid, which destroys not only RNA, but also mucopolysaccharides. This led him to conclude that they contained a carbohydrate protein complex, and he called them mucoprotein grana. But since the cytological investigations could not give complete proof of their chemical nature, he turned to the press sap.

His protein preparations showed significantly higher carbohydrate content when obtained from hardened than from unhardened plants. They also gave a positive test with phloroglucin, showing the presence of pentoses or pentosans. Glucosamine or other aminosugars were not found, therefore the substances could not be mucoproteins. He concludes that they must be glycoproteins which can bind water strongly due to their many hydrophilic groups—much more strongly than can pure proteins. He even suggests that the often observed rise in osmotic value of the cell sap on hardening is an adaptation to prevent injury to the protoplasm due to excessive swelling of the glycoproteins. The possible nature of the carbohydrate was indicated by the similarity in the spectroscopic absorption between extracts and D-ribose solutions. Further evidence was the increase in hardiness obtained by treatment with D-ribose as compared with glucose, though the tests for hardiness have already been criticized (see above).

Some of JEREMIAS' results do not agree with his conclusions. For instance, the grana were more numerous in hardened plants of two varieties but not in others. In the case of the protein preparations, the carbohydrate content was higher in January samples from the open than in May samples. The May samples themselves showed higher carbohydrate contents than the greenhouse material, though undoubtedly no more hardy. Nor was it possible to distinguish the hardiness of different varieties by the carbohydrate content of the protein complex. This agrees with the negative results reported above by LEVITT (1954).

The Russian workers have recently resurrected the old idea that hardiness is associated with lipids, which are accumulated at the protoplast surface. But PIENIAZEK and WISNIEWSKA (1954) point to the long known fact that at least one of their tests is invalid. They failed to find any such lipid accumulation with any of the tests used.

Earlier workers attempted, unsuccessfully, to show some relation between enzyme activity and frost resistance (LEVITT 1956). Even the few recent investigations have failed to throw light on the problem. CHRISTOPHERSEN and PRECHT (1956) have shown that the temperature at which yeast is cultured affects the rate of CO_2 production under anaerobic conditions as well as dehydrase activity. But the relation is unexpected, for the "cold resistance" of these enzyme systems increases with rising culture temperature.

Protoplasmic factors have received some attention. It is interesting to note that even when grown in the laboratory in lighted thermostats, *Lemna minor* shows seasonal changes characteristic of plants undergoing hardening—slow growth, high osmotic value, and high permeability (to uric acid) in winter, the reverse in summer (PIRSON and GOELLNER 1953). In the case of mulberry cortical cells (SAKAI 1955 b), though the permeability to water rose from September to a maximum in December, there was no subsequent decrease, and the permeability of the cells in spring could not be used as a criterion of frost hardiness. In agreement with earlier results by SIMINOVITCH on trees (see LEVITT 1956), JOHANSSON et al. (1955) have found a conversion of water insoluble to water soluble protein in wheat during hardening. This was reversed during dehardening. The same parallelism between water soluble protein and hardiness occurred in the two hardy alfalfa varieties tested by BULA et al. (1956); though the least hardy variety failed to show any relation between the two. HODGSON and BULA (1956) found 100–200% increase in water soluble protein N per gram fresh weight of sweetclover roots during a natural hardening period from Sept. 20th to Oct. 18th in Alaska. Unfortunately, non-protein N showed just as much of an increase; and since results were expressed on a fresh weight basis some of the change must have been due to the decrease in water content that normally occurs during hardening. The overall N increase also indicates a net transfer of nitrogen from the leaves or the soil or both. Such results therefore fail to prove a relation between water soluble protein and hardiness.

Many workers are still attempting to show a relation between protoplasmic viscositity and hardiness, in spite of the fact that the commonly used methods (centrifugation and plasmolysis) are usually not applicable to the problem (LEVITT 1956). Thus, the centrifuge method frequently fails because of differences between the density of both the particles and the cytoplasm in hardy and non-hardy plants. This is partly due to a greater dehydration of the cytoplasm in hardy cells by the more concentrated cell sap. As a result of this greater dehydration, a higher cytoplasmic viscosity would be expected in the hardy cells, and this has been found in wheat cells (JOHANSSON et al. 1955, JEREMIAS 1956). The Russian workers have also reported an increased viscosity in winter on the basis of centrifuge tests, but PIENIAZEK and WISNIEWSKA (1954) were unable to find any significant difference. JEREMIAS admits that it is difficult to obtain quantitative results in wheat cells by either the centrifuge or plasmolysis time method, but he states that it is certainly higher in the hardened. He also found that the press sap of hardened plants was definitely more viscous than that of the unhardened, since it took longer to centrifuge down particles. This has long been known (LEVITT 1956) but is simply explainable by the higher sugar content. He also found that extracts of hardened plants increased the protoplasmic viscosity of *Helodea canadensis,* extracts of unhardened plants decreased it as compared with water controls. But there was no difference related to the hardiness of different varieties. It is difficult to see how this can give any information related to hardiness, since *Helodea canadensis* is incapable of hardening. It is even possible from his results that the increase is due to an injurious geling of the protoplasm. When the most reliable methods are used, no consistent differences in cytoplasmic viscosity can be detected between hardy and non-hardy cells in their normally hydrated states (see Levitt 1956). But it is when the two kinds of cells are at the same state of dehydration that this value is important, and under such conditions the hardy cells possess a lower viscosity (Levitt 1956). It should be pointed out, however, that in all these cases it is structural rather than true viscosity that is measured.

Scarth and Levitt's (1937) discovery that hardy cells show convex plasmolysis, non-hardy concave plasmolysis, has been confirmed by the Russian workers (e. g. Henkel) though interpreted in a different way (see Pieniazek and Wisniewska 1954). They concluded that the convex plasmolysis is due to absorption of plasmodesmata into the protoplasm during the winter rest, so that the protoplasm loses contact with the cell wall. In this state, according to them, it can no longer be injured by ice crystals in the intercellular spaces. Siminovitch and Levitt (1941) had, indeed, previously shown by photomicrographs that the convex plasmolyzed cortical cells of *Hydrangea* fail to reveal cytoplasmic strands when weakly plasmolyzed, in contrast to similarly plasmolyzed non-hardy cells. But when these same cells were plasmolyzed more strongly, the strands became visible. This was interpreted to indicate that the strands of the hardy cells were so highly hydrated when slightly plasmolyzed that their refractive index was essentially that of the surrounding solution, and they were invisible. Only when more strongly dehydrated, was their refractive index high enough for visibility. Pieniazek and Wisniewska (1954), however, were unable to confirm these results in the case of apple, pear, and cherry shoots. By use of phase contrast microscopy, they confirmed the explanation of convex plasmolysis as a plasmodesm absorption. They did not find complete absorption as stated by the Russian workers (e. g. Oknina), but were able to determine the degree of frost resistance by counting the cells with convex plasmolysis. This method worked well with five apple varieties but a sixth variety failed to show any difference between winter and spring behavoir. The differences between these results await further investigation. It may be that some species or varieties behave differently from others. Perhaps the plasmolysis used was still not strong enough to observe the highly hydrated strands. It is still more likely that the results depend on the plasmolyte used, and that they can be explained on the basis of protoplasmic consistency and adhesion. The cytoplasmic strands may be present at incipient plasmolysis, but due to the high protoplasmic mobility they may be able to flow back into the protoplast on further plasmolysis when sugar solutions are used. In salt solutions, though the strands are still highly hydrated and mobile, they may adhere more tenaciously to the cell wall. Certain cations have long been known to favor such adhesion, and all affect the physical properties of the protoplasm in some way or another.

Recent observations by Asahina (1956) fully agree with the explanation of the role of cell permeability in hardiness suggested by Scarth and Levitt (1937). In non-hardy plants, the rate of development of surface ice crystals soon became very slow unless the cell was irreversibly injured. This is to be expected from their low permeability. Soon after the growth of these crystals ceased, intracellular freezing occurred even with slow cooling. In the case of hardy cells, on the contrary, the surface ice crystals grew very large, continuously withdrawing water from the cells. This was presumably due to the high permeability of the hardy cells which permitted the water to exosmose rapidly to the external ice. The hardy cells underwent remarkable dehydration and contraction, as a rule, and none froze intracellularly.

Pseudoplasmolysis was observed after extracellular thawing. This was not always fatal, e. g. in the case of hardy cortical cells of woody plants and semihardy herbaceous plants. This again agrees with theory, for in these cells with high permeability the pseudoplasmolysis would be slight and short-lived. In fact, the hardy cells usually failed to show any pseudoplasmolysis at all after thawing. Non-hardy cells could be killed by even a slight degree of freezing, with or without pseudoplasmolysis.

Asahina believes that his observations of freezing and thawing support the theory that when extracellular freezing occurs, there is irreversible coagulation due to freezing rather than thawing. As an example, isolated watermelon cells frozen in oil occasionally froze extracellularly and contracted strongly. On warming, they gradually expanded with thawing, and when the surrounding ice was almost melted, the protoplasmic surface frequently exhibited a finely granular frothy appearance. At an instant sometimes several minutes after thawing, an irregular and sudden partial expansion occurred as if the ectoplast had partially ruptured, resulting in fatal collapse of the protoplast, with or without any apparent release of the cell contents. Thus, though death occurred on thawing, the injury to the surface was apparent well before thawing was complete.

5. Frost Resistance in Animals

Unlike drought and heat resistance, the problem of frost resistance is basically the same in plants and animals. Yet it has not been as intensively investigated in animals as in plants, at least until recently. The earlier literature is reviewed by Luyet and Gehenio (1938) and will not be dealt with here.

A. Avoidance

All homoiotherms, of course, possess frost avoidance as long as they can maintain their normal body temperatures at environmental temperatures below freezing. As a rule, they can stand little or no freezing of their tissues and therefore are not frost tolerant. But this is true even of poikilotherms. Scholander et al. (1953) showed that the frost tolerance of arctic fish is much less than usually believed. According to some reports blackfish (*Dallia pectoralis*) can be frozen solid in winter without injury. Scholander et al., however, showed that the actual freezing of the tissues causes irreversible damage just as in the case of homoiotherms, though if only parts are frozen the fish may still survive.

Insects seem to have been investigated more thoroughly than any other group of animals. They may show both frost avoidance and tolerance. In the majority of cases, those that survive winter without injury are devoid of frost tolerance. According to Salt (1950), "Although the few species which hibernate in very exposed situations can survive the formation of ice in their tissues, the remainder are killed if they become frozen. The latter depend for their survival on the insulating protection of their hibernacula and on their ability to undercool." He found it difficult, in fact, to eliminate undercooling in insects. Many normally undercooled 20 to 30° C. and some considerably more for at least a short time. It is their ability to undercool that Salt (1956 b) calls "cold hardiness."

Table 9. *Percentage of Cephus cintus larvae that froze when exposed to — 15° C. Larvae intact in their stubs which were covered with moderately moist soil.*

(From Salt 1950).

Exposure period (days)	frozen (%)
100	78
110	88
120	82
130	97
140	96
150	98
160	98
200	100

Nevertheless, he (1950) concluded that the undercooling points are unreliable by themselves as a measure of insect cold hardiness. Insects held in the undercooled state were found to freeze at irregular intervals, often after long periods of time. This freezing is, of course, initiated by the formation of an ice crystal nucleus. The probability of formation of this nucleus is dependent on (1) the extent of undercooling, (2) the cold hardiness of the insect, and (3) the length of time undercooled. Therefore, unless the population is adequately protected by its environment and by a high degree of cold hardiness, losses from freezing will take place by degrees during the entire winter (Table 9).

Asahina et al. (1954) agree that cold hardiness of insects is chiefly a matter of prevention of freezing. In fact, they showed a clear periodicity in the undercooling points. These were high in summer, dropped in autumn, reached a minimum during winter, and rose again in spring (Table 10). Salt (1956 b) concluded that supercooling is constant over a wide range of moisture conditions. Only when desiccation was severe did it produce appreciable cold hardening. Chilling at a consistent low temperature increased the cold hardiness (i. e. lowered the undercooling point) in one species but was ineffective or doubtfully effective in others. On the other hand, the variable temperatures of the insect's natural environment produced a significant cold hardening in all four species.

Besides this gradual hardening of the non-feeding insect. which is due to an ecological factor he (1953) distinguishes an abrupt change due to change of stage or cessation of feeding. This would no doubt, be his explanation for the periodicity in hardiness described above by Asahina et al.. since these changes in hardiness were associated with a change in stage.

He is also skeptical about the commonly stated relationship between hardening and reduced moisture content, since much of the earlier evidence is based on a comparison of feeding and non-feeding insects.

The importance of food in cold hardiness of insects is ascribed by SALT (1953) to the greater ease with which it freezes in the digestive tract and inoculates the insect's tissues. In this way, undercooling of the insect is prevented or at least reduced, and cold hardiness is markedly decreased since ice formation in these insects results in death. Thus feeding larvae were much less cold-hardy than non-feeding premoult or freshly moulted larvae. Similarly, foreign bodies implanted in the insect decreased the undercooling and therefore the cold hardiness of hardy larvae and pupae.

B. Tolerance

As mentioned above, some insects do show frost tolerance. SCHOLANDER et al. (1953) point out that overwintering larvae of arctic insects freeze brittle during winter. Chironomid larvae spent the winter frozen in the first inches of the muddy bottom of a shallow pond. These larvae were usually about 10 % dehydrated during winter, i. e. the water content of their tissues was 60—70%. Yet they survived temperatures of —20° C. for months and —50° C. or lower for shorter periods. In artificial tests, the larvae froze rapidly at — 6 to — 10° C., changing from transparent red to completely opaque yellowish orange—suggesting that they were filled with highly refractive ice crystals. With decreasing temperature they became more brittle. Even at — 32° C., however, they were still somewhat pasty, and the broken surface yielded a fine drop of unfrozen, dark "blood" when squeezed. They could be frozen and thawed repeatedly without injury. The amount of ice formed at different temperatures was measured dilatometrically (Fig. 5). Since the curves were the same during freezing and thawing, a true equilibrium must have occurred with no supercooling of any of the parts. The same curves were obtained even with previously boiled harvae. Similar freezing curves have been obtained by SALT (1955, 1956 a).

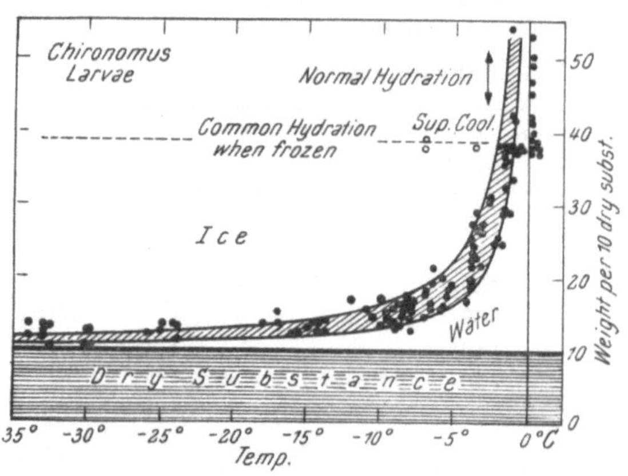

Fig. 5. Amounts of ice and water in *Chironomus* larvae at different freezing temperatures. (From SCHOLANDER et al. 1953.)

ASAHINA et al. (1954) found that the "slug caterpillar" has a blood freezing point of about — 2° C. yet the prepupae very readily supercool. When cooled below — 20° C., they suddenly freeze hard (Fig. 6) and may withstand this freezing for as long as 100 days without any injury. Even the next generation was normal. Daily freezing and thawing could also be repeated many times without injury. During the warm season, however, they could not survive freezing at —10° C. for a few days. Definite seasonal changes in frost tolerance were observed (Table 10). Judging from both the freezing curves and direct observation of the tissues, the process was found to be as follows. At first the blood freezes very rapidly, then extracellular freezing of the tissue cells occurs. These tissue cells are therefore dehydrated and contract. When washed with hypotonic salt solution (0.15 M NaCl) before freezing, the tissues were easily frozen intracellularly and viability was lost on thawing. Similarly, when active summer larvae were subjected to freezing at temperatures lower than —10° C., the tissues usually froze intracellularly.

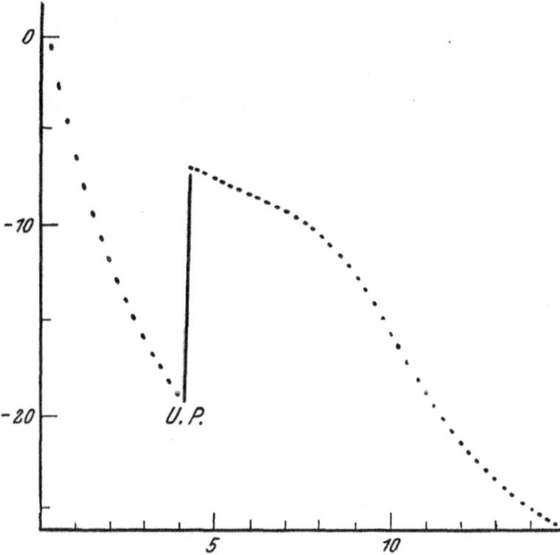

Fig. 6. Freezing curve for an overwintering prepupa. Ordinate = temperature (° C), abscissa = time (mins.) (From ASAHINA et al. 1954.)

Other animals besides insects, have also been shown to possess tolerance. KANWISHER (1955) froze molluscs, echinoderms, annelids, and arthropods which occur only below the tide level and are therefore not exposed to freezing under natural conditions. None of these were able to survive the freezing. On the other hand, intertidal forms were found to withstand the freezing of as much as 75% of their body water without injury; and this occurred at temperatures that they were regularly exposed to twice daily in nature (e. g. —20° C.). The actual interior temperatures of these

Table 10. *Seasonal variations in frost-resistance of larvae (Cnidocampa flavescens).* (From ASAHINA et al. 1954.)

Stage	Date	Undercooling point (°C.)	Final body temperature (°C.)	Duration of freezing (min.)	Results
Feeding larva	5. IX.	— 4.2	— 4.0	8	alive
Feeding larva	8. IX.	— 4.7	— 5.8	11	dead
Two weeks after spinning	10. IX.	—10.2	— 4.8	6	alive
Two weeks after spinning	10. IX.	—15.3	—17.7	11	dead
Prepupa	24. X.	—15.9	—15.0	16	alive
Prepupa	19. XII.	—24.0	—31.7	10	alive
Prepupa	28. IV.	—20.7	—28.7	16	alive
Prepupa	2. V.	—17.0	—9.3	5	alive
Prepupa	20. V.	—	—10.0[1]	300	alive
A few days before pupation	12. VI.	—	—5.0[1]	15	dead

[1] Air temperature near the insect.

invertebrates were found to be within a few tenths of a degree of air temperature. Some were able to supercool to —5° C., but in no case to —7° C.

The greatest stimulus to research on frost tolerance in animals has come from the discovery that spermatozoa and red blood cells may be subjected to severe freezing without injury. Under suitable conditions, many other animal cells have also survived this treatment. The early literature as well as many of LUYET's own classical researches on the subject is reviewed by LUYET and GEHENIO (1938). In most cases, however, success depended

Table 11. *Hemolysis of red blood cells on freezing and thawing at different rates.* (From LOVELOCK 1953 a.)

Time between −3 and −40° C. (sec.)		Hemolysis (%)
Freezing	Thawing	
0.3	1.5	7
0.3	50	90
3.0	2	10
2.8	24	52
9.0	2	35
9.0	43	93

on ultrarapid freezing and thawing in order to avoid microscopically visible ice crystal formation. The water was thus retained in the cells in the "vitreous" (non-crystalline) state, or as LUYET (1954) later explained it, the crystals were submicroscopic in size.

The critical temperature region for red blood cells is —4 to —40° C. (LOVELOCK 1953 b). This was proved by ultrarapid freezing to —160° C. in a bath of dichlorodifluoromethane followed by transfer to an alcohol bath at —60° C. which was heated at a rate of 1°/min. Tubes removed at —41° C. showed 20% hemolysis; those removed at —39° C. and similarly thawed showed 80% hemolysis. Within the critical temperature region, the amount of hemolysis varied with the speed of freezing and thawing (Table 11). It is interesting to note that in the case of the bacterium *Pasteurella Tulariensis,* the critical region is —30 to —45° C. (MAZUR et al. 1957).

From a practical point of view ultrarapid freezing is of limited importance, since the necessary speed can be obtained only when the specific surface of the tissue or suspension is very great. Furthermore, in some cases survival requires slower freezing (see SMITH 1954). Consequently, it was the later discovery that the inclusion of certain solutes in the medium prevented injury at slower rates of freezing and thawing that gave the greatest stimulus to research (see SMITH 1954).

The most effective solute so far is glycerol (see SMITH et al. 1951). In the presence of this solute, for instance, spermatozoa survived freezing at —79° C., though only if the freezing was "slow" (i. e. cooled to —79° C., over a period of 5—40 mins. Such cells survive storage at —79° C. for a whole year (SMITH 1954). In order to find out just what was happening during freezing and thawing, SMITH et al. observed the process under the microscope. In the absence of glycerol, spermatozoa were compressed into narrow clefts between the ice crystals; in the presence of glycerol, smaller crystals formed at lower temperatures and there was no suggestion that crystals ever contained spermatozoa. Similar results were obtained with red blood cells. 10 or 20% glycerol was required to prevent injury. Crystals were never seen within the cells, though due to the smallness of the cells they could not be certain of this observation and therefore could not eliminate the possibility that glycerol protected by preventing intracellular ice formation. They did conclude, however, that the formation and dissolution of crystals in the medium obviously caused less mechanical damage to the cells in the presence of glycerol than in its absence. Red blood cells did not become crenated when frozen in the presence of glycerol, due perhaps to exchange of equal volumes of glycerol and water through the cell membrane.

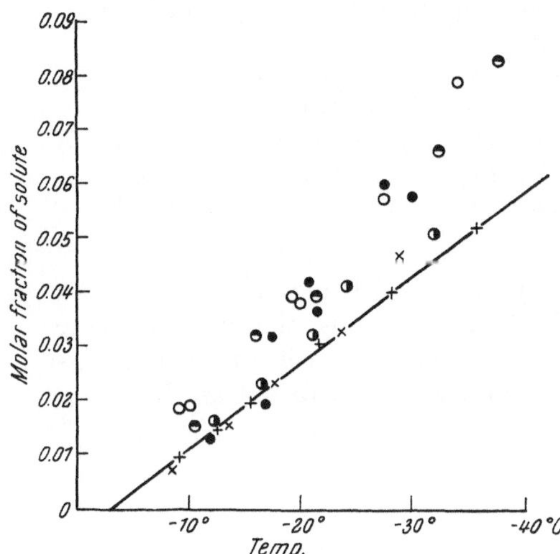

Fig. 7. Relation between solute concentration (in 0.16 M NaCl) and temperature to which red blood cells can be frozen without hemolysis. ○ = methanol and ethanol, ◓ = formamide and acetamide, ● = ethylene and propylene glycol, + = glycerol, x = di- and triethylene glycols, ◑ = monoacetin. (From LOVELOCK 1954.)

One interesting observation was a more rapid deterioration of the frozen cells after thawing than in the case of freshly collected material. This agrees with the observation of post-thawing injury to plant cells (LEVITT 1957 a). LOVELOCK (1954 b) tested 15 neutral solutes and found that protection occurred only when the solute was able to penetrate the cell. Those that penetrated poorly (e. g. xylose, glucose, erythritol) gave slight protection. Those that penetrated well (glycerol, ethylene glycol, diethylene glycol) gave complete protection if sufficient was present (2.5 M). No protection was afforded by the non-penetrating sucrose and polyethylene glycol. Furthermore, when penetrating solutes were used, there was a direct relation between the solute concentration and the temperature to which the red blood cells could be frozen in 0.16 M NaCl without hemolysis (Fig. 7). In the case of bacteria, however, even hypotonic solutions of glucose and lactose permit recovery from freezing at —75° C., though freezing in saline at the same temperature causes complete killing (MAZUR et al. 1957).

This protective effect of glycerol has led to some controversy as to the mechanism of frost injury. LOVELOCK (1953 a, b and 1954 a) and MERYMAN (1956) explain the injury by the high salt concentration that results on ice-formation, and the glycerol protection by the reduction in this concentration. HARRISON (1956) has recently attempted to apply the same explanation to the frost killing of bacteria. In agreement with this concept, survival was far better in distilled water than in broth (Table 12). This

Table 12. *Survival of Escherichia coli in different concentrations of solute at — 22° C.*
(From HARRISON 1956.)

Menstruum	Viable cell count per ml. after storage for		
Initial viable cell count per ml.: 1.4×10^8			
	1 day	3 days	1 week
Undiluted broth	1.4×10^7	1.9×10^6	3.2×10^5
Broth diluted 10-fold	3.6×10^7	9.8×10^6	4.5×10^6
Broth diluted 100-fold	4.0×10^7	3.2×10^7	3.7×10^7
Distilled water	1.2×10^8	9.0×10^7	9.6×10^7

is. of course, the opposite of what would be expected if mechanical crushing of the cells occurred. As in the case of the animal cells, the bacteria were largely frost killed at —22° C. if in NaCl, but glycerol prevented this injury. As further evidence in favor of solute concentration as the cause of injury, HARRISON cycled bacterial suspensions between concentrated solute and diluent without freezing, and obtained the same series of pendent curves (with steadily increasing injury) as in the case of repeated freezings and thawings. Glycerol markedly decreased the slope of the curves.

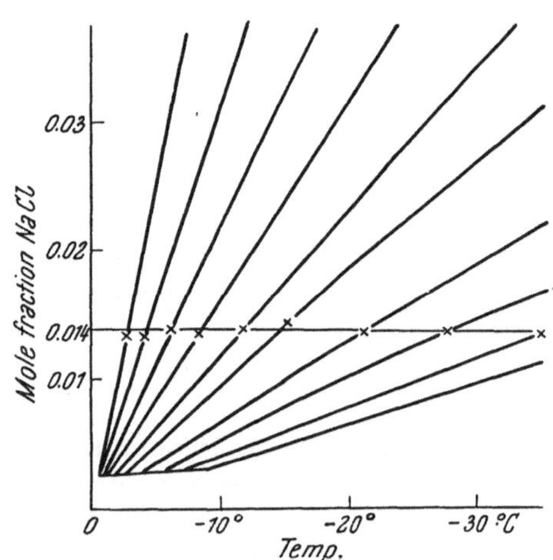

Fig. 8. Red blood cells frozen in 0.16 M NaCl containing increasing quantities of glycerol from 0.0 (1st curve) to 3.0 M (last curve) In each case hemolysis first occurs (points marked by ×) at a temperature resulting in a NaCl mole fraction of 0.014 (0.8 M). (From LOVELOCK 1953 b.)

LOVELOCK's theory is based on some striking quantitative data. He (1953 b), for instance, showed that when red blood cells are frozen at different temperatures in 0.16 M NaCl containing glycerol in different concentrations, the killing temperature varied inversely with the glycerol concentrations. Yet in each case hemolysis occurred at 0.8 M NaCl (Fig. 8). This indicates rather conclusively that injury depends on the NaCl concentration. But is does not depend solely on exposure of the cells to concentrated NaCl, since this alone failed to reproduce the damage obtained during freezing (Fig. 9). If, however, exposure to strong salt solution was followed by immersion in 0.15 M NaCl, the injury obtained resembled that from freezing and thawing (Table 13). This fact, in itself, shows that the frost injury cannot be solely due to the high concentration of NaCl during freezing.

Yet LOVELOCK explains the effect as follows: "The nature of the destructive action of concentrated salt solutions appears to be connected with their ability to dissolve or disperse essential constituents of the cell membrane, in particular lipids and lipoproteins." This agrees with the above experiments in which frost injury always occurred at an *external* NaCl concentration of 0.8 M. But it cannot explain the fact that the injury is obtained only on transfer to a weak NaCl solution. Furthermore, in another experiment he showed that the glycerol produces a much greater protection against frost injury when it penetrates than when it does not. Yet whether or not the glycerol penetrates the cell, the NaCl outside the cell (in contact with the membrane) must attain the same concentration at any one freezing temperature. Consequently, frost injury occurs at a lower salt concentration in contact with the cell membrane when the glycerol fails to penetrate than when it does. His own results therefore, disprove his "lipid dissolving" theory. It also must follow that the concentration of NaCl attained inside the cell would be the same whether or not the glycerol penetrates and again the salt concentration theory would not apply; unless the cell becomes less permeable to the NaCl as well as to the glycerol. But this reduction in permeability would lead to more injury when the glycerol penetrates than when it does not, on the basis of the salt concentration theory. Consequently, LOVELOCK's results disprove his lipid dissolving theory, as far as the NaCl concentration is concerned.

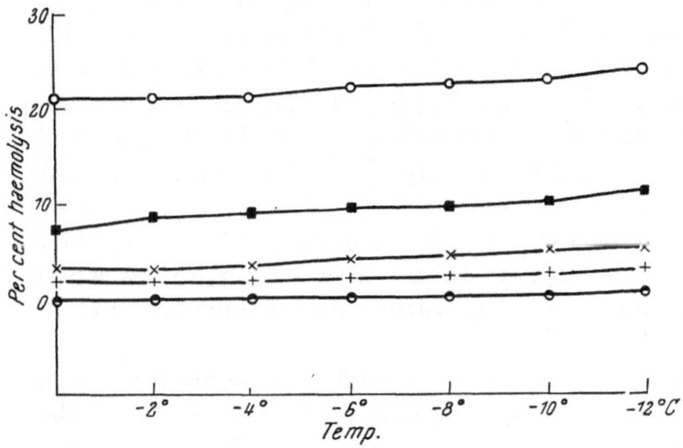

Fig. 9. Hemolysis of red blood cells after 5 minutes in NaCl solutions (0.15 M bottom curve, to 4.0 M top curve) at different subzero temperatures.
(From LOVELOCK 1953 a.)

Table 13. *Percentage of hemolysis of red blood cells resuspended in 0.15 M NaCl after 5 min. exposure to strong NaCl solutions.*
(From LOVELOCK 1953 a.)

Temp. (°C.)	Conc. NaCl solution (molar) 1.0	1.5	2.0	2.5	3.0	4.0
20	2.7	11.3	38	74	79	81
0	3.7	13.3	41	81	80	81
−5	3.5	12.8	41	80	80	81

Lovelock (1953 b) apparently realizes these facts for he points out that "The fact that glycerol must be present within the cell before it can exert its protective action suggests that the internal KCl concentration is at least as important as the external NaCl concentration in causing the damage." This again does not agree with his results, for in that case the addition of non-penetrating glycerol should actually increase the frost injury, since it must increase the dehydration of the cell (at least before freezing equilibrium) and therefore the internal concentration of KCl. In actual fact, he obtained some protection by use of non-penetrating glycerol. Thus, cells suspended in non-penetrating 2.5 M glycerol containing 0.16 M NaCl showed no more hemolysis as a result of freezing at — 15° C. than did cells in 0.16 M NaCl alone at about — 5° C. Furthermore, if salt concentration were the direct cause of injury, why should "snap freezing and rapid freezing" seldom give good survival (Smith 1954)? In further disagreement is the fact that destruction occurs rapidly during the first 3 minutes after freezing, then slows considerably (Lovelock and Polge 1954). Some hemolysis occurred even within a few seconds at — 3 to — 40° C. Furthermore, in the case of bacteria, for which Harrison (1956) has applied the same explanation of frost injury, he found little or no injury when frozen in distilled water at the same temperature as produced killing in broth or saline.

The salt concentration theory is therefore incapable of explaining frost injury in animal cells. The theory that has proved most successful in the case of plant cells is a mechanical theory of injury. This is not the same as the older theory of injury due to the pressure of ice, and that Lovelock opposes. It explains frost injury on the basis of the stresses that arise in the protoplasm as a result of the cell contraction and subsequent expansion during extracellular freezing and thawing. This theory is in accord with all the known facts that have accumulated from work with plant cells (Levitt 1956).

But it might seem impossible to apply the same theory of frost injury to plants and animals, since the two react in diametrically opposite ways to external non-toxic solutions. Plant cells show maximum protection against frost injury when the solute does not penetrate, little protection when it does (Levitt 1956). Animal cells, on the contrary, show maximum protection against frost injury when the solute does penetrate, none when it does not. How can this difference be reconciled?

In the case of plant cells, the protection against frost injury by non-penetrating solutes was shown to be dependent on plasmolysis (Levitt 1956). In this way, the cell wall remained in its normal position regardless of the amount of ice formed in the intercellular spaces and therefore the tension on the protoplasm (that otherwise accompanied the collapse of the cell as a whole) was eliminated. Animal cells in general do not plasmolyse. Therefore they cannot be protected from frost injury by non-penetrating solutes. In fact, these solutes, if hypertonic, will produce the very same kind of injury as the frost itself since they result in the same kind of cell collapse due to the osmotic removal of water.

In the case of animal cells, the protection against frost injury by penetrating solutes is readily explained by the much less severe cell collapse on freezing, due to the increased osmotic value of the cell sap. What about plant cells? The older experiments failed to result in more than a very slight increase in frost resistance. But this was due to the fact that very frost sensitive plants were used—plants normally completely killed by —2 to —3° C. (LEVITT 1956). Failure to protect these plants is to be expected, since the osmotic protection is proportional to the original frost resistance. When plants that are already somewhat frost resistant are used, very significant increases in resistance result from treatment with penetrating solutes—increases that are precisely what would be expected from a purely osmotic effect (LEVITT 1957 b).

Certain other basic differences exist besides the lack of plasmolysis in animal cells. Unlike plant cells, the red blood cells and spermatozoa are bathed in a salt solution. Furthermore, the cell walls of these animal cells are, no doubt, less rigid than those of mature plant cells. In spite of these basic differences between plant and animal cells, LOVELOCK's results can be explained by ILJIN's mechanical theory of frost injury. The fact that frost injury always occurs when the NaCl concentration outside the red blood cells reaches 0.8 M, led him to conclude that this concentration of NaCl is the cause of the injury. But even if this is true, it does not follow that the high concentration of salts is the direct cause of injury. It is just as logical to ascribe the injury to the large osmotic removal of water. In fact his method of duplicating the frost injury by immersing in strong NaCl and then transferring to weak NaCl is essentially the same as the plasmolysis and deplasmolysis method of measuring frost hardiness in plants that has proved uniformly successful (LEVITT 1956).

Since the cells were originally in 0.16 M NaCl, an increase in concentration to 0.8 M NaCl would mean that the cells must have lost 80% of their water, provided that the NaCl did not penetrate appreciably during the 10-minute freezing period. This would also agree with the greater injury by Li than by Na salts (LOVELOCK 1953 a), since the Li^+ is more highly hydrated. Equimolar solutions would therefore result in greater dehydration than NaCl. Similarly, the fact that 0.08 M Na_2SO_4 is as injurious as 0.15 M NaCl would be explained by a combination of the greater number of ions per molecule and the greater hydration of the $SO_4^=$ than the Cl^-. The same explanation would hold for the injury obtained in the absence of freezing when the red blood cells were immersed in hypertonic NaCl. In these experiments, LOVELOCK found that the glycerol concentration could be varied from 0.0 to 5.2 M without affecting the amount of hemolysis. This is unexpected if the injury is due to the amount of water removed. However, the experiment was performed in such a way as to prevent the protective effect of glycerol. The cells were not pretreated for 10 minutes at 40° C. (as was done in the freezing experiments to permit penetration of the glycerol). Consequently, the glycerol would at least temporarily dehydrate the cells even more than the NaCl alone and would therefore temporarily increase the damage until the glycerol penetrated.

In agreement with the mechanical theory of frost injury, LOVELOCK (1953 b) points out that in NaCl solutions between 0.8 M and 2.0 M, the blood cells are sensitive to mechanical shock, sudden chilling, and to transfer back to isotonic media. This is precisely what has been found for plant cells (LEVITT 1956). The last point is analogous to the deplasmolysis injury that can actually be used as a measure of frost resistance in plants (LEVITT 1956). Since the blood cells are sensitive to mechanical shock, they must also be sensitive to the stresses of frost dehydration. The injurious effects of "snap freezing" and the rapid progress of the injury (facts that cannot be explained on the basis of a salt concentration effect) are just what would be expected from such mechanical injury. This is why the glycerol must penetrate the blood cells in order to produce the protective effect. Only in this way can the stresses of frost dehydration be prevented. In the case of plant cells, protection is more easily obtained by non-penetrating solutes which plasmolyse the cells and completely prevent the stresses.

The differences between the nature of frost injury in plant and animal cells may, therefore, be more apparent than real, and the same mechanical theory of injury may be applied to both. That the external solution in which the cells are frozen may sometimes prove toxic was long ago discovered by MAXIMOV for plant cells (see LEVITT 1956). But this is a totally different phenomenon and cannot explain the normal frost injury that occurs in the absence of such toxic substances.

6. Drought Resistance

As a rule, so-called xerophytes are drought resistant. But a tremendous variety of plants are included in this group (KILLIAN and LEMÉE 1956) and some are exceptions to the rule. Ephemerals, for instance, do not give evidence of drought resistance since they complete their life cycle during the brief moist period and are not exposed to drought during their vegetative period. Xeromorphy has even been ascribed to N-deficiency in cold soils, though GREB (1957) was able to separate the two and to show that xeromorphy is primarily due to water deficit. On the other hand, many plants that possess drought resistance do not show xeromorphy and would not be classified as xerophytes. The resistance may be due to avoidance or tolerance.

A. Avoidance

Consideration of drought avoidance separately from drought tolerance is advisable not only because different factors are involved, but also since, as mentioned above, the two often seem to be mutually exclusive. This was shown by PARKER (1951) in the case of conifers. The differences in drought resistance between species were due to avoidance and not to tolerance. The drought resistance was judged from field observation—*Ponderosa* pine is often found in pure stands on comparatively dry sites, Douglas fir on more moist sites, and Western arborvitae on even more moist slopes. Rates of water loss from detached leaves or branches were generally in the reverse order. Yet in all cases, death occurred when leaf moisture dropped to below about 50% of the dry weight. It is unfortunate that the tetrazolium test was used for the latter determinations, since this method is not always unequivocal. Later (1954), however, even the method of plasmolysis resistance (see LEVITT 1956) failed to reveal any difference in tolerance between the conifers tested. Similar results have been obtained by CLARK and LEVITT (1956). Soybeans that are subjected to a "hardening" drought develop a reduced rate of water loss (Fig. 10) but fail to increase in drought tolerance.

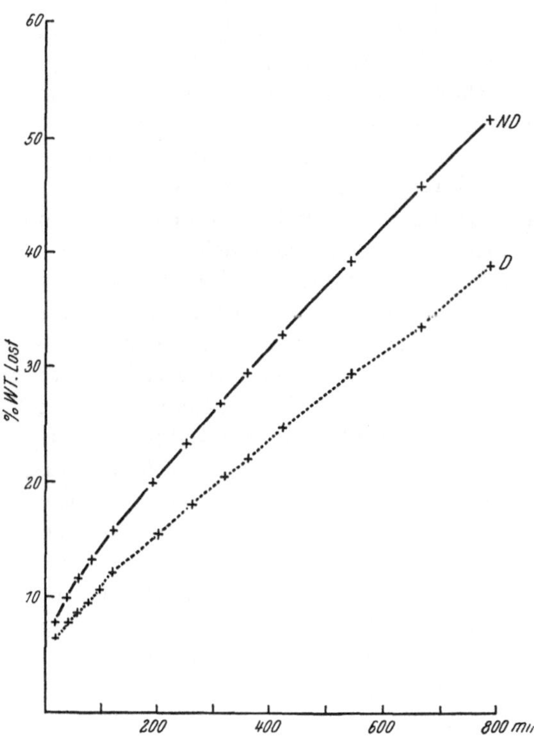

Fig. 10. Water loss (% of fresh wt.) by excised shoots of droughted (*D*) and non-droughted (*ND*) soybean plants. (From CLARK and LEVITT 1956.)

SLATYER (1955) has clearly shown that the superior drought resistance of grain sorghum over cotton and peanut is due to avoidance. He measured the "relative turgor"

$$\left(\frac{\text{actual water content}}{\text{water content when turgid}}\right)$$

of the leaves and found that with the onset of dry conditions, grain sorghum maintained the highest turgor while cotton showed the most rapid drop of the three. As a result, growth rate in cotton was reduced as soon as a slight soil moisture stress appeared; in grain sorghum it was not reduced until the stress was severe. Peanut was intermediate between the two. That two factors were involved was indicated by the two stage decline in relative turgor. SLATYER logically explained this by an initial lag of absorption behind transpiration due to a rapid rise in the latter. The later decrease in turgor would then be due to a reduction in absorption. This term "relative turgor" is an unfortunate one since it is not directly related to turgor pressure which, in fact, drops to zero after loss of very little water. Most of these "relative turgor" values would therefore correspond to zero turgor pressure. A much better quantity for measuring the reduction in water content of the plant is the "natural saturation deficit" as opposed to the "critical saturation deficit" (see LEVITT 1956). These have now been adopted as standard terms by the German and Austrian schools of workers as a result of OPPENHEIMER's (1932) pioneering experiments. SLATYER's results point clearly to the two main factors in drought avoidance: resistance to water loss and increased water uptake.

1. Resistance to water loss

The tremendous range in transpiration among plants extends all the way from a daily loss of $^1/_{4300}$ of the plant's weight to an hourly loss of five times its weight (Killian ind Lemée 1956), the one being about 500,000 times the other. Recognition of the importance of this factor has swung back and forth from one extreme to the other. It was first proposed by Schimper as the prime factor in drought resistance (Maximov 1929). But Maximov (1931) emphasized that this concept was based largly on a teleological point of view, and that the only evidence was indirect, based on observations of morphological adaptations. When, however, investigators actually measured transpiration rates, they soon found that the xerophytic plants usually had the highest rates (Maximov 1929). Even then, however, Maximov (1931) cautioned that "It is not the rate of transpiration when an abundance of water supply is present, but the capacity to restrict water loss to a minimum in time of drought that characterizes the water utilization of the xerophyte." Schratz (1931) came to the same conclusion.

Nevertheless, Maximov (1931) insisted that though different types of xerophytes possess different structural modifications, it is a fact that under desiccating conditions the changes are always in the same direction—decrease in cell size, an increased number of stomata per unit area, a more compact network of veins, denser hair covering, thicker cuticle and wax covering, and often an increased development of palisade tissue. These adaptations, according to him, all favor increased assimilation and transpiration. Ferri (1955) recently reported that these xeromorphic characters are much more developed in a less arid than in a more arid Brazilian desert.

It is now apparent that the actual relation between transpiration rate and drought resistance cannot be described without indicating the degree of hydration of the plant. Huber (1931) showed this clearly by allowing shade and sun shoots to become saturated before determining their transpiration rates, then cutting off their water supply. The shade leaves lost water more rapidly at first, then the sun leaves, then the shade leaves again (Fig. 11). After 15 hrs., the shade shoots had lost 44%, the sun shoots 58% of their respective original water contents. Yet this condition was later reversed and the sun shoots finally transpired only half as rapidly, though their water content was now higher than that of the shade shoots. As a result, the sun shoots became air dry two days later than the shade shoots. These last changes do not show in the figure because the rates of water loss are too small for the scale used (less than 1 mg/g. f. w./hr.). The resistance to water loss may be due to the following three factors: (a) stomatal closure, (b) low cuticular transpiration, (c) reduction in leaf surface.

a) Stomatal closure. Huber (1931) found that the most drought resistant species of those he studied (*Teucrium montanum*) had a greatly reduced transpiration rate during a drought that reduced the soil moisture to a suction force of 100 atms. But even then, the stomata were open in the earliest morning hours, and this species was the last to close its stomata in the morning (at 7–8 o'clock). This would permit photosynthesis to continue for a longer time than in the other less drought resistant species.

It has been amply shown that different species behave very differently in this respect. OPPENHEIMER (1953) observed a pronounced drop in transpiration to very low levels, with complete cessation in certain cases during the hot hours of the day. This was due to stomatal closure in the case of two sclerophyllous species (*Quercus calliprinos* and *Laurus nobilis*). *Phillyrea media,* however, behaved quite differently. Its transpiration persisted at an even level throughout the dry summer, and the stomata were always wide open. This resulted in the development of an extremely

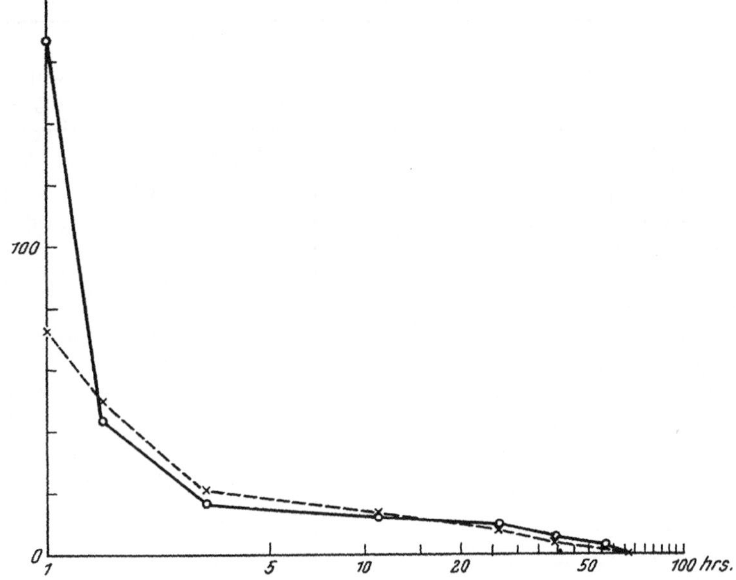

Fig. 11. Transpiration rate (mg./g. f. w./hr.) of excised shade branches (solid line) and sun branches (broken line) of *Quercus pedunculata.* Time in hours on logarithmic scale. (From HUBER 1931.)

high osmotic potential, as determined cryoscopically, i. e. a very marked dehydration.

Perhaps the most striking contrast between these two types is provided by FERRI's (1955) measurements in two different dry regions of Brazil. In one of these regions (the "cerrado") water is stored in the soil equivalent to 3 years' rainfall. Most plants of this region, though exposed to very marked atmospheric drought, show no restriction of water expenditure either at the beginning or the end of the dry season. The much more arid region (the "caatinga") has no water reserves in the soil throughout the year. The rivers dry up during the 7-month dry season. Even when rain does occur, the run-off is heavy. The average temperature is very high, the relative humidity low, and strong winds persist all day and evening. Plants growing in this extreme drought markedly restrict their water loss—even during the rainy season. Their stomatal action is much more rapid than in the case of cerrado plants (Table 14 and Fig. 12). An extreme example is *Spondias tuberosa,* whose stomata may close completely within 5 minutes and are open only for a few hours during the less severe morning.

According to STOCKER (1956) the more resistant of two oat varieties showed a greater degree of stomatal closure at noon when the maximum drought occurred. SHMUELI (1953) found a 25% decrease in average daily stomatal opening of the banana when the soil moisture decreased.

b) Low cuticular transpiration. According to PISEK and WINKLER (1953), conifers can withstand a little more dehydration (about 8% of the

Table 14. *Rate of stomatal movement and cuticular transpiration.* (From FERRI 1955.)

Species	% Reduction in initial transp. rate	Time (mins.)	Cuticular transpiration (% of total)
a) Cerrado:			
Byrsonima coccolobifolia	50	30	15
Didymopanax vinosum	50	20	
Stryphnodendron barbatimam	50	20	
Kielmeyera coriacea	45	50	
Erythroxylum tortuosum	40	50	
Erythroxylum suberosum	20	50	
Anona coriacea	60	30	2
Andira humilis	40	20	33
Palicourea rigida			35
b) Caatinga:			
Spondias tuberosa	55	2	2—5
Caesalpinia pyramidalis	50	10	15
Jatropha phylacantha	50	2	10—20
Maytenus rigida	50	10	15
Ziziphus joazeiro	50	5	15
Bumelia sartorum	50	3	10
Aspidosperma pyrifolium			15
Tabebuia caraiba			15
Capparis yco			10

saturation water content) before beginning to close their stomata, than broad-leaved trees (about 3%). But in both cases closure is about complete at a dehydration of 10–17%, after which cuticular transpiration is the controlling factor. The lower the cuticular transpiration, the longer the leaves were likely to withstand drought. It has long been known that some drought resistant plants have very low rates of cuticular transpiration. The classical example is the cactus that holds onto a large fraction of its water even if deprived of all contact with water for a year or more. Similarly, the more drought resistant sun leaves have been found to transpire more rapidly than the shade leaves on the same plant; yet when cut from the twigs, their transpiration drops to a lower value (STOCKER 1929). PARKER (1951) was able to show that the relative drought resistance of several conifer species, judged by their habitats, parallels the resistance to loss of water from their leaves. SATOO (1956) also found that the order of cuticular transpiration (both in the case of intact plants and in excised shoots) in three conifer species was inversely related to the order of their drought resistance. He concluded that this was one of the two most important factors, the other being the development of the root system (see below).

MAXIMOV (1931) points out that the cuticular transpiration is only 2–5 times lower than stomatal in mesophytes, but as much as 25 times

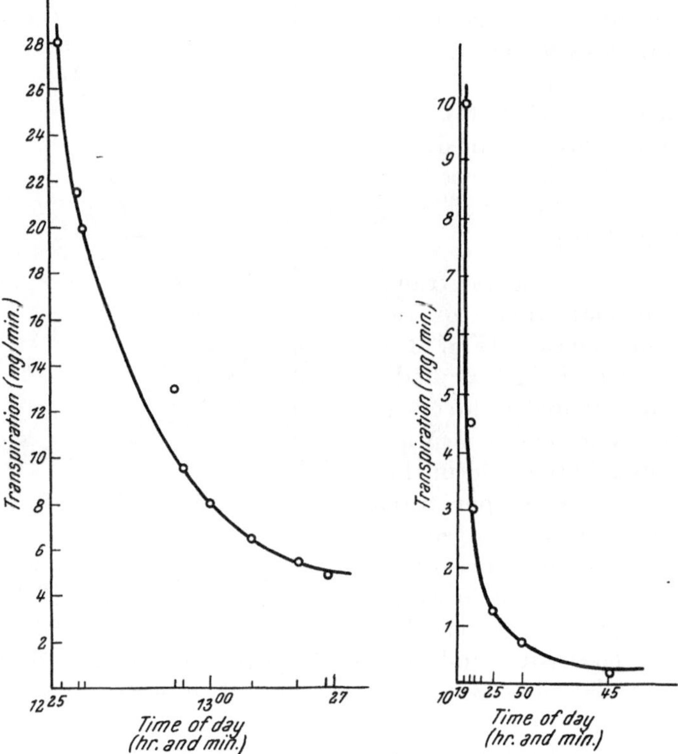

Fig. 12. Rate of transpiration and of stomatal closure in the cerrado plant *Stryphnodendron barbatimam* (left) and in the caatinga plant, *Spondias tuberosa* (right).
(From FERRI 1955.)

lower in the case of leaves with thicker cuticle. FERRI (1955) obtained experimental evidence corroborating this for the plants of the most arid Brazilian area (the "caatinga") but not for those of the somewhat less arid "cerrada" (Fig. 12). In the caatinga, the transpiration was actually less during the dry month of January than in the moist month of April, the former being mainly cuticular, the latter largely stomatal. In this case, in striking contrast to the plants of the cerrada, transpiration rate and evaporation curves form mirror images of each other (Fig. 13). The various leaf modifications responsible for low cuticular transpiration have been amply discussed and described (see SHIELDS 1950, PARKER 1956).

It has recently been shown (Fig. 10) that even the relatively mesophytic soybean plants may be able to reduce cuticular transpiration as a result of exposure to drought (CLARK and LEVITT 1956), i. e. they can undergo a pseudohardening. It is not a true hardening for there is no increase in drought tolerance. The only detectable change is an increase in leaf surface lipids in proportion to the decrease in cuticular transpiration (Table 15). It was even possible to bring the transpiration rate back to that of moist grown plants by removing some of the leaf surface lipids (Fig. 14). The reverse treatment—spray deposits on foliage has recently been used to improve survival of longleaf pine plantings (ALLEN 1955).

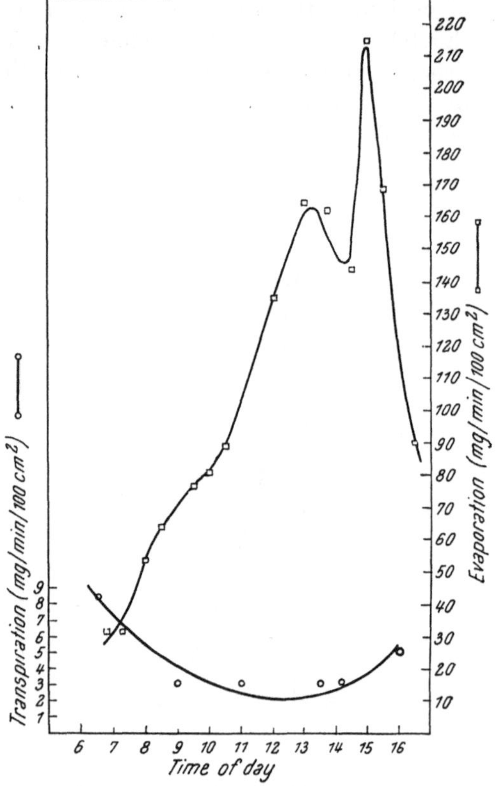

Fig. 13. Diurnal course of evaporation (Piche evaporimeter) and of transpiration from *Jatropha phyllacantha*. (From FERRI 1955.)

Both stomatal and cuticular transpiration are proportional to the incident radiant energy (Fig. 15). Consequently, some of the leaf modifications may conceivably reduce water loss by reducing heat absorption from the sun, since they could in this way keep both their temperatures and their vapor pressures from rising. By keeping their vapor pressures low, they would, of course, prevent an increased water loss. A possible case in point is SCHIEFERSTEIN and LOOMIS' (1956) recent statement that "there is no clear-cut correlation between surface wax and xeromorphic adaptation,

Table 15. *Transpiration loss and lipids removed per unit area of excised leaves from droughted (D) and nondroughted (ND) soybeans.* (From CLARK and LEVITT 1956.)

Exp. Nr.	Transpiration loss in grams for 1½ hrs (starting ½ hr. after excission)			$\frac{D}{ND} \times \frac{(1.236)\ \text{AREA ND}}{\text{AREA D}}$	Lipids per cm² leaf surface		
	ND	D	D/ND		ND	D	ND/D
1	.107	.052	.486		.0174	.0273	.637
2	.110	.061	.554		.0000	.0123	
3	.155	.074	.477		.0233	.0434	.536
4	.071	.035	.493		.0222	.0325	.638
5	.154	.068	.441		.0206	.0291	.707
a. v.	.1194	.058	.486	*.600*	.0167	.0289	.577

so that the presence or absence of wax appears to be of doubtful survival value." The main basis for their conclusion is the assumption that the only way in which leaf surface wax could aid drought survival is as a barrier to water loss. And since it does not form a continuous layer it cannot be an effective barrier. But its effect on water loss may conceivably be indirect. If it succeeds in reflecting a large amount of the incident radiation, this may result in a lower leaf temperature and therefore a lower vapor pressure. The water loss could, in this way, be markedly affected. The problem can be solved by means of a direct determination of drought resistance or simply of transpiration rate before and after removal of the wax. The above suggested protective mechanism against rise in temperature will be considered in more detail under heat resistance.

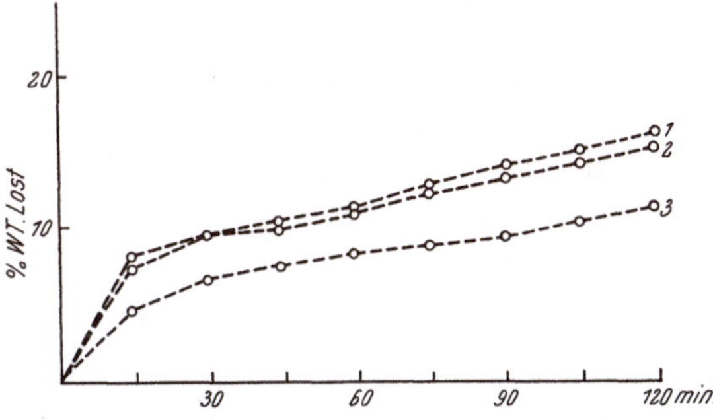

Fig. 14. Increased transpiration by droughted soybean leaves after removal of some surface lipids (curves 1 and 2). Control leaf curve 3.

(From Clark and Levitt 1956.)

c) Reduction in leaf surface. In spite of the generally lower cuticular transpiration of succulents, their water loss per unit area may sometimes be quite high (Kilian and Lemée 1956). In these cases, the principal adaptation responsible for reduced water loss is the reduction in their specific surface. Rolling, folding, or shedding of leaves have long been

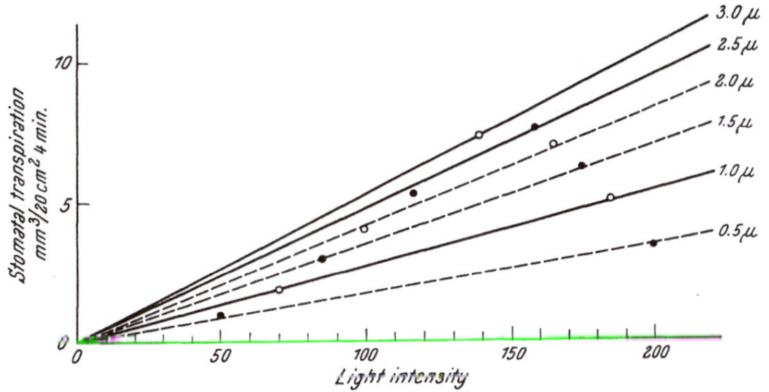

Fig. 15. Relationship between light intensity and transpiration by *Phaseolus vulgaris* at various stomatal apertures.

(From Brouwer 1956.)

recognized as aids to drought resistance (Maximov 1929, Shields 1951). Clements (1937 a, 1937 b) could find no response to drought in the case of tomatoes and sunflowers other than in shedding of the lower leaves. Ferri (1953) gives a striking example of leaf folding in dry weather (Fig. 16). In the caatinga, when the drought reaches its maximum severity the leaves drop and the vegetation becomes whitish. This is responsible for the name of the region which means "white forest." Only a few of the most resistant species are able to retain their foliage during this period.

As pointed out above, avoidance of drought injury simply involves prevention of a reduction in the plant's water content to the point of injury. An understanding of the mechanism of avoidance therefore does not require any understanding of the mechanism of drought injury, provided it is a direct injury due to reduced water content. If, however, the very mechanism of avoidance introduces a secondary kind of injury, then this must be considered as a factor in drought adaptations.

Avoidance due to a reduction in water loss does just this; for the most efficient method the plant has for reducing water loss to a minimum (short of shedding its leaves) is to keep its stomata closed. But this brings net photosynthesis to a stop. Consequently, a drought resistant plant that depends on reduced water loss for its resistance must possess a mechanism for keeping photosynthesis going. This explains the above observation of HUBER's (1931) that the most drought resistant species not only reduced its transpiration rate to a minimum during drought, but opened its stomata in the earliest morning hours and was the last to close them at this time of least danger of water loss. According to STOCKER (1956) the more resistant of two oat varieties does the same thing, for it opens its stomata more rapidly in the early morning when the drought is at a minimum. The extremely drought resistant succulents have an additional mechanism. Their respiratory breakdown at night does not go to completion, but results, instead, in the accumulation of organic acids (VICKERY 1953). Recent investigations have also shown that they keep their stomata open at night and fix CO_2 in the dark (THOMAS 1951) again forming organic acids. These intermediate organic acids can apparently be later photosynthesized to starch during daylight (VICKERS 1953).

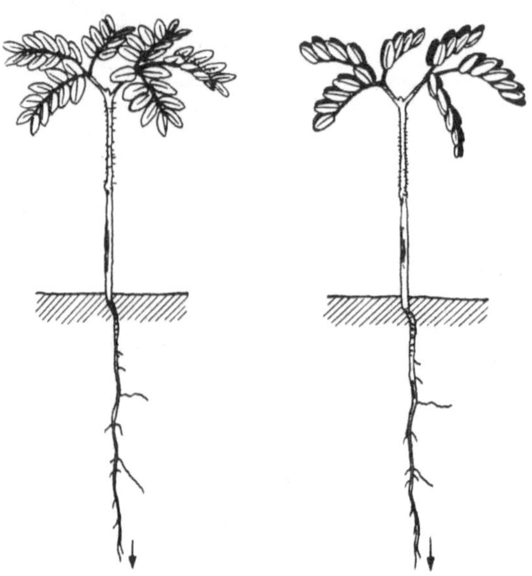

Fig. 16. Seedlings of *Caesalpina pyramidalis* with leaves folded under dry conditions (right), leaves open in high humidity (left.)
(From FERRI 1953.)

But these adaptations merely make the best of a bad situation. They cannot lead to the most rapid rates of synthesis and growth. In fact, ECKHARDT (1952) found that, in general, those species possessing the greatest ability to reduce the total transpiration in summer also have the most markedly reduced CO_2 absorption. Consequently, the most prodigal plants may actually be the most efficient producers of dry matter—provided that they have developed methods of increasing their water uptake. It is not surprising that these "water spenders" can displace the "water savers" provided that enough water can be found in the soil to squander.

2. Increased water uptake

a) From the soil. Those xerophytes that transpire more rapidly than mesophytes under similar conditions without losing their turgor must, of course, be absorbing water more rapidly. It is obviously these efficient water absorbers that Maximov (1931) was most concerned with, and he felt that this adaptation is far more common and important than the ability to reduce water loss. Schopmeyer (1939) adopts this concept to explain the greater drought resistance of shortleaf than of loblolly pine. The former was able to maintain a higher water content than loblolly even when soil moisture was limited; yet it transpired more rapidly than loblolly, and the solute concentration and bound water of the two was the same (Fig. 17). Raheja's (1951) high correlation between relative transpiration rate and the ratio of foliage to root system in a number of varieties of crop plants during drought points to the importance of the root system.

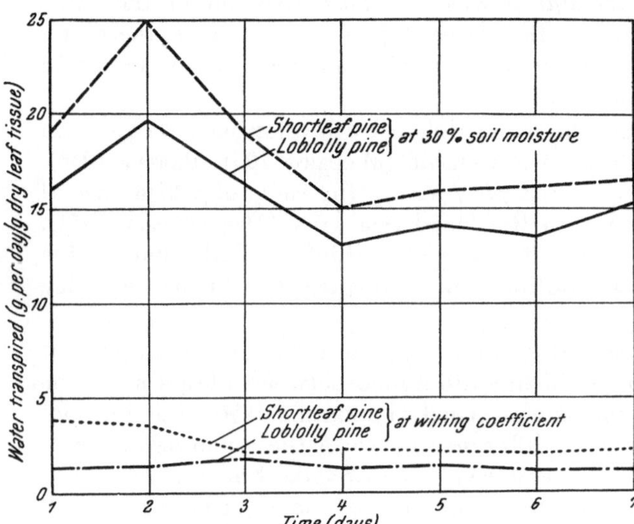

Fig. 17. Daily transpiration of loblolly pine and shortleaf pine at 30% soil moisture and at wilting coefficient. Water content of shortleaf remained higher in spite of higher transpiration. (From Schopmeyer 1939.)

Maximov's concept has also suffered from the pendulum of opinion. Though at first considered as a logical explanation of drought resistance, it was opposed by Briggs and Shantz' classical discovery that the wilting coefficient of soils is essentially the same for all plants—drought sensitive as well as drought resistant. These results have been disputed from time to time, but now appear to be more firmly established than ever (Veihmeyer 1956).

But does this fact dispose of the possibility that different plants may remove different quantities of water from the soil? Obviously, the plant with the larger root system will explore a larger volume of soil and therefore will have removed more total water when the wilting coefficient is reached (Kramer and Coile 1940—see Parker 1956.) That drought increases the ratio of root to top has long been known (Maximov 1929). Simonis (1952) has even found that if this increase in root system is taken into account, the total dry weight of plants grown under dry conditions may actually exceed that of similar ones grown with full moisture, though the above ground part is considerably reduced in size. Satoo (1956) has shown that this factor may lead to a reversal in the relative drought resistance of conifer species. *Cryptomeria japonica* was more resistant than *Chamaecyparis obtusa* when the roots were permitted to penetrate freely into deeper layers of soil. This was associated with a lower top: root ratio in the former. When, however, the depth of root penetration was limited by planting in shallow containers, the order of resistance was reversed. On the other hand, *Pinus densiflora* was the most resistant of the three under both conditions, and it had the lowest top: root ratio. It is, therefore, obvious that extent of root system may or may not be a major factor in the drought resistance of conifers.

Some plants may overcome the water deficit by developing a root system that is deep enough to reach the water table (Maximov 1931, Parker 1956). An extreme example of this is *Prosopis farcata,* the roots of which extend down 15 m. to the water table in the Dead Sea area (Oppenheimer 1951). According to Maximov (1931) this is one of the chief methods that enable plants to continue their assimilation and transpiration during hot, dry periods.

There is still a third possibility that may at times be an important factor. Though there is no significant difference between plants as far as the wilting coefficient of a soil is concerned (Veihmeyer 1956) this does not mean that there is no significant difference between the amounts of water removed by different plants from equal volumes of soil. For it is inconceivable that the wilting coefficient for a soil can be *exactly* constant for all plants, even though constant within the limits of error of the method used to measure it. That would be equivalent to saying that two leaves with relative humidities of 98.5% and 99% have cells at the same degree of hydration, because relative humidity measurements cannot be made that accurately. In this case, if the values are correct, there is a difference of more than 6 atms. between the cells of the two leaves. In the same way, Killian and Lemée (1956) found a suction tension of 18–20 atms. in the roots of one of the Gramineae in the Sahara desert at the wilting coefficient, whereas in other annuals it was less than 14.5 atms.

Just how accurate measurements of wilting coefficients are, is difficult to say. Veihmeyer (1956) concludes that they are constant to within ½–1%. In his table 2, for instance, he considers the difference between 10.0 and 10.5% not significant. Suppose, then, these were the *exact* values for plants A and B. Plant A would be able to remove 5 ml. more water than plant B per kg. soil. If these are two corn plants of different varieties, we might expect them to have root systems 1 m. in diameter and 2 m. deep (Miller 1931). They would, therefore, each explore approximately 1.5 m^3 or 1500 l. soil. Assuming a soil specific gravity of about 2, this would weigh 3000 kg. Consequently, plant A would remove 15 liters more water from the soil than plant B, due to only a 5% difference in wilting coefficients. This is 7% of the total water lost by a corn plant in a four month growing period (MILLER 1931), i. e. equivalent to the total loss of about six average days. This can hardly be called an insignificant quantity, and could certainly mean the difference between life and death of plants exposed to drought. But this factor seems to be quantitatively less important than the other two.

b) From the air. It has long been known that mosses absorb water through their leaves. ANDERSON and BOURDEAU (1955) have in fact concluded that fog, dew, and mostly rain are the only effective sources of water for mosses. According to their results, liquid water can be conducted only 1–2 cm. up the stem of *Polytrichum commune* and *Atrichum angustatum* in sufficient quantity to produce turgidity in the leaves. But even in the case of higher plants, much has been made in recent years of the long known fact that water can be absorbed by leaves from a saturated atmosphere or one containing fog (see STONE et al. 1956 for a review of the literature). Many concluded from this that some mysterious force is involved and that the pressure developed by the absorption of water through leaves is greater than when water is absorbed through roots. When, for instance, the leaves were in contact with a fog, water was "pumped back" through the roots into an empty flask (BREAZEALE et al. 1950). Similarly, when dry soil was collared around the stem of a tomato plant growing in well watered soil, the moisture content in the collar was built up only to the wilting point (BREAZEALE and MCGEORGE 1949); but when the plant absorbed water through its leaves from a saturated atmosphere, the build up of water in the collar reached field capacity or higher (BREAZEALE et al. 1950).

The above misinterpretation of these results appears to be a part of a recent trend to reject the well established, physical laws of diffusion, and to attempt to show that water absorption by plants is an "active" process. Just as such attempts have been fruitless with respect to water absorption by roots and tubers (LEVITT 1953, ORDIN et al. 1956), so, too, the above attempts with leaves have proved to be in vain. As a result of more careful control of experimental conditions, HAINES (1952, 1953) and STONE et al. (1956) have clearly shown that water movement from leaves, back through the plant and out through the roots is not due to any "pumping" action, but is a simple diffusion process. It occurs only when temperature fluctuations lead to distillation of the water. When the temperature is kept constant, no such backward water movement into the flask occurs. Consequently, if plants survive as a result of dew or fog, it is simply due to diffusion of water into the above ground parts.

Water absorption by above ground parts of the plant has been well established under certain conditions, and may conceivably be a factor in drought resistance. The analogy may be drawn with desert succulents that owe part of their drought resistance to a shallow, spreading root system that quickly absorbs the moisture from light rains (MAXIMOV 1929). Small amounts of moisture deposited on the leaves may be equally important in some cases, for as MAXIMOV (1931) points out, the very life of desert plants depends on their ability to profit from more or less short and comparatively favorable periods of moisture supply. Differences in such abilities may readily exist between plants, for as ARVIDSSON (1951) points out, the amount of dew deposited on a leaf will vary inversely with the leaf temperature. Her measurements showed a drop of 2.8° C. below that of the atmosphere in some species, only 0.8° C. in others. In general, thin leaves with long margins (e. g. grasses) cooled the most. The variation in the amount of dew formed on different plants is shown in Table 16.

Table 16. *Absorption of dew through leaf surfaces.*
(From ARVIDSSON 1951.)

Species	Average amount of dew (% of fresh wt.)	Average amount of dew absorbed	
		% of fr.wt.	% of dew
Plantago major	25.3	3.0	10.0
Taraxacum vulgare	33.5	1.2	3.2
Polygonum aviculare	16.0	1.9	11.7
Ranunculus acris	25.8	1.8	7.3
Trifolium pratense	40.1	1.8	3.6

Sun leaves with thick cuticle failed to condense water; shade leaves with thin cuticle did. In general, when the natural water deficit of the leaves was slight, the dew was absorbed only to the extent of 1–3% of the saturation weight. Larger amounts were absorbed when the deficits were

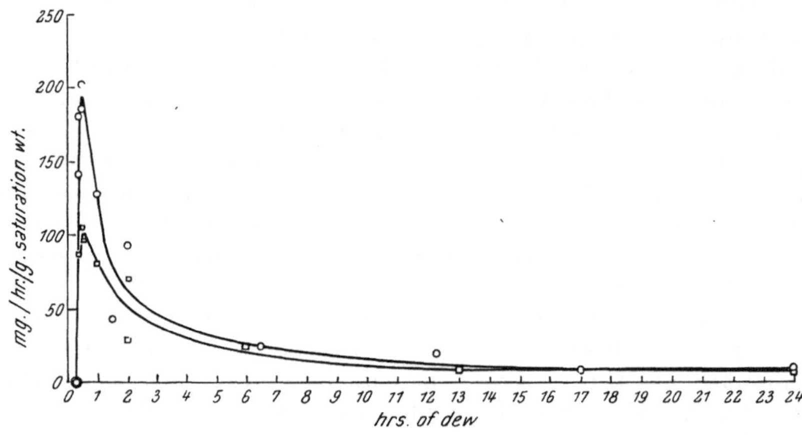

Fig. 18. Speed of absorption of artificial dew by barley leaves with 40% (○) and 20% (□) deficit respectively. (From ARVIDSSON 1951.)

high. This was clearly shown when barley was exposed to artificial dew (Fig. 18). Light rains were also readily absorbed by the species investigated, to the extent of 7–15% of the saturation weight, reducing the deficit by as much as 50%. The evidence also indicated absorption of fog by leaves. STEUBING (1955), in general, confirmed ARVIDSSON's main results, though the actual quantities obtained differed somewhat. The dew absorbed by the 17 species investigated varied from 1.7 to 13.1% of the fresh weight of the leaves. Since she was working in a climate characterized by larger saturation deficits than the Swedish island investigated by ARVIDSSON, it is not surprising that the absorbed dew produced a greater reduction in the leaf deficits.

c) Improved translocation of absorbed water. Among the several tree species that HUBER (1931) studied, the most drought sensitive (*Fagus, Picea, Tilia*) showed the most difficulty in supplying water to their upper parts. This was indicated by the marked reduction in transpiration rates of their leaves with increased height (Fig. 19). In the more drought resistant

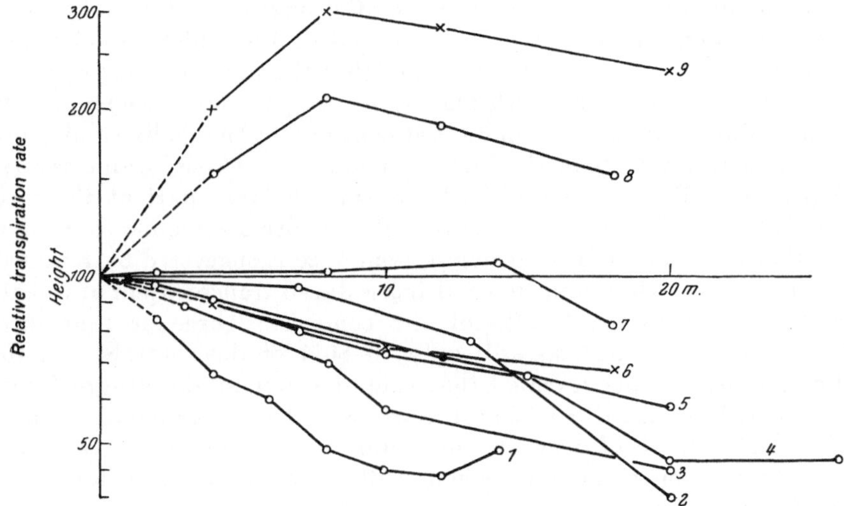

Fig. 19. Transpiration rates of some trees at different heights (meters) *1. Sequoia gigantea*, *2. Fagus sylvatica*, *3. Tilia grandifolia*. *4. Picea excelsa*, *5. Acer pseudoplatanus*, *6. Quercus sessiliflora*, *7. Fraxinus excelsior*, *8. Larix leptolepis*, *9. Pinus austriaca*. (From HUBER 1931.)

species (*Quercus* and *Pinus*), there was an actual increase with height. HUBER therefore agreed with MAXIMOV in assigning a minor role in drought resistance to reduced transpiration. The frictional resistance to water movement was further shown by cutting off shoots and putting them in water. This doubled the transpiration rate of shade shoots, though that of the sun shoots rose insignificantly. Further evidence of the large frictional resistance to water movement in a mesophyte was produced by TENNANT (1954). Even at a relative humidity of 88%, more water was lost by shoots of the moss *Thamnium alopecurum* than could be taken up from a potometer. THODAY (1931) concluded that the major significance of reduction in leaf size is the reduced frictional resistance to flow of water. This follows from the shorter distances of the mesophyll cells from the main channels of supply. OPPENHEIMER (1949) found the rate of transpiration of oaks related to the width of vessels, in agreement with the earlier results of HUBER and ROUSCHAL (see OPPENHEIMER 1951).

It is now obvious that there is no real basis for controversy as to the relation between water exchange and drought resistance. As in the case of so many other scientific controversies, this one stemmed from the reluctance to admit that though $5+5=10$, $6+4$ can also equal the same sum. There does seem to be some relation, however, between the degree of drought and the water exchange. In general, drought resistant plants lose water more rapidly during moderate droughts than do the less resistant mesophytes. This is due to their greater ability to obtain water from the soil and to translocate it to the transpiring surfaces. At more extreme droughts, the drought resistant plants lose water less rapidly than the less resistant plants, mainly because of better stomatal control and lower cuticular transpiration. These differences hold not only when comparing species with different maximum degrees of drought resistance, but also when the same individual plants are compared in different states of resistance. There is no longer any question but that the most drought resistant of all plants may not possess such high rates of water loss. These readily reduce their water loss to a minimum yet continue to metabolize and grow, though at a slower rate than the "water spenders." These "water savers" are adapted to soil drought that lasts for such a large part of the year that even the most efficient absorbers of soil moisture are unable to survive.

One other possible factor in drought avoidance is suggested by KRAMER's (1950) observations. He found that wilting reduced transpiration of tomato and sunflower plants to 10–40% of the controls. Three or four days after rewatering, the rate was still only 70–80% of the controls. At the same time, measurements revealed that rate of water intake dropped 50% in plants wilted overnight, 80–90% in those wilted for four days. Tomato showed a high degree of recovery but sunflower failed to recover much of its water absorbing capacity when well watered for four days. He explains these results mainly by a cessation of elongation and an increased suberization, both of which would reduce the area of the root surface freely permeable to water. But the recovery of a plant's ability to absorb water must be closely related to the actual drought injury to the roots during wilting and therefore is clearly a case of drought tolerance.

B. Tolerance

This property should be measured by determining the threshold relative humidity, rather than the critical water content, that causes injury. Clear evidence of this has recently been produced by SATOO (1956), who found the former value closely related to drought resistance in three conifers while the latter showed no relationship to drought resistance.

Drought tolerance is a more stable property in some species than in others. Among some 30 species of bacteria and fungi, BURCIK (1950) found the threshold relative humidities to be remarkably constant for each species. The lowest values were 90% for bacteria, 88% for yeasts, though only a few members of each group attained these low values. He was unable to adapt the organisms to lower relative humidities even over periods of 6 months. In the case of mosses and liverworts (ABEL 1956 a and b), on the other hand, this threshold value may be highly variable both for different cells of the same organ and for all the cells depending on whether or not they are predried at 96% rel. hum. The latter result depends on the nature of the species (Table 17). It is interesting to note that the "hygrophytes" among the mosses survive a lower relative humidity (82%) than the above xerophytes among bacteria and yeasts.

Table 17. *Variations in threshold relative humidity of untreated and predried (at 96% rel. hum.) bryophytes.*

(From ABEL 1956 a.)

Group	Untreaded ("primary" threshold)	Predried	Amplitude
Hygrophytes	low (82%)	low	0 — small
Xerophytes	high	high	small — 0
Mesophytes..............	low–medium	high	large

As shown above, in the case of higher plants that are mainly homoiohydric (WALTER 1950), the most extreme droughts, particularly if long continued, are usually survived due to avoidance. As a result, their water deficits under natural conditions are small and fall within a relatively narrow range that is well below the critical saturation deficit (Fig. 1). Even in the case of these, however, desiccation resistance (or tolerance) may be in the same order as drought resistance, as SATOO (1956) found to be true of three conifer species. But this relation held only when the root system was restricted. In other words, it is possible for a drought resistant species to overcome a lack of drought tolerance by the development of one or more drought avoidance factors.

On the other hand, all the lower plants that WALTER classifies as poikilohydric possess maximum drought tolerance and minimum avoidance. Plants of this type do not have the capacity to grow and metabolize during drought, but they are capable of becoming air dry without injury, and can grow and metabolize actively again when the drought is over. Some higher plants have also developed this capacity to a greater or less degree and their natural water deficits cover a wider range and reach higher values than in the case of the drought avoiding plants under the same conditions (Fig. 1). It is obvious that all gradations and combinations occur in different species. The actual water deficit tolerated usually increases from early to late summer (Fig. 20).

The most extreme case of drought tolerance in higher plants is the creosote bush (DUISBERG 1952). In spite of its mesophytic morphology, it is the dominant plant of the hottest and driest plains and deserts of Mexico and southwest United States. Under conditions of extreme drought the more mature leaves die; but the immature leaves and buds dry out and turn brown. When more favorable moisture conditions exist, the immature leaves and buds continue growth. These drought hardy immature leaves have lower contents of nordihydroguaiaretic acid and resin and a higher protein content than the mature, non-hardy leaves; but it seems hardly likely that these differences can explain the difference in drought resistance.

In contrast to the rather consistent increase in frost tolerance produced by added solutes, the effect seems much more variable in the case of drought tolerance. Among the several species of bacteria and fungi tested by Burcik (1950), addition of salt solution to the substrate raised the threshold relative humidity for 4 species, lowered it for 2 species, failed to change it for 5 species. Among the alga and two fungi tested by Füchtbauer

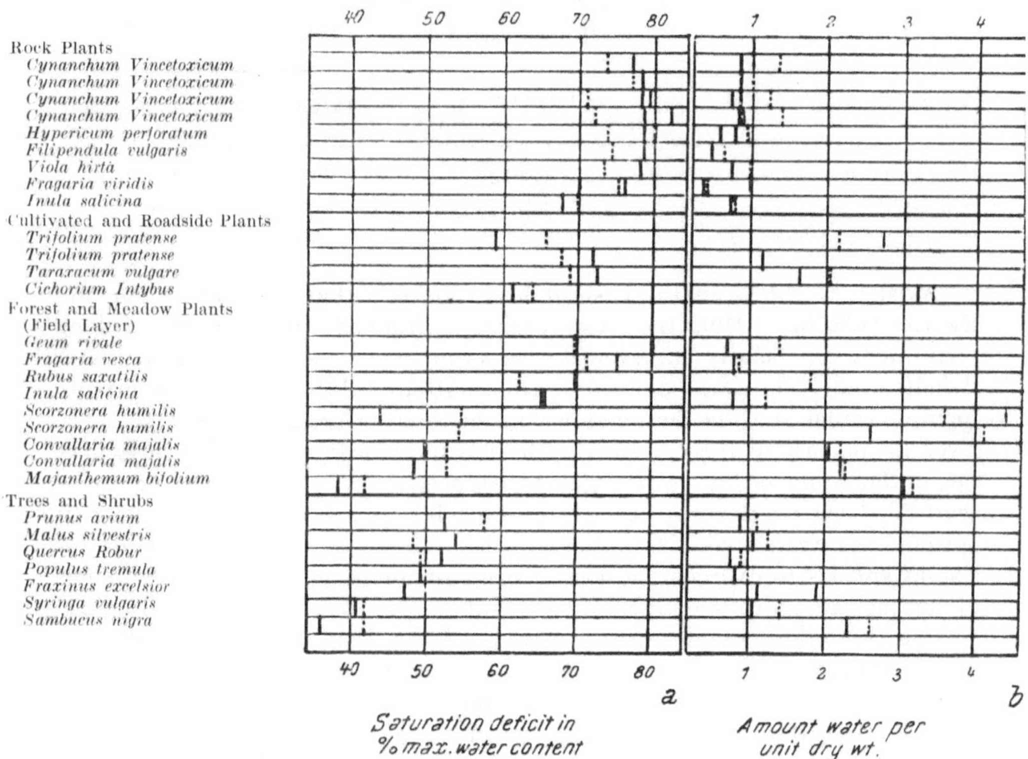

Fig. 20. Early (broken line) and late (solid line) summer average critical saturation deficits (left) and water per unit dry matter at the critical saturation deficit (right). (From Arvidsson 1951.)

(1957 a, b) only baker's yeast and the alga (*Chlorella pyrenoidosa*) showed a significant increase in tolerance with the addition of sorbitol to the medium. The increased tolerance was not due to any purely physical effect of the sorbitol, since the organisms had to be grown in its presence before developing resistance (Table 18). Nor did the sorbitol have to be present in the surrounding solution during the drying process. Thus, after growth in the presence of sorbitol, the cultures could be washed with NaCl or other solutions or even with distilled water and left there for 3 hrs. before drying without loss of the increased drought tolerance (Table 19).

In his second paper (1957 b) Füchtbauer attempted to determine whether this increased drought resistance was due to a change in "bound" or "retention" water. The method he used was essentially similar to that employed by TODD and LEVITT (1951) and by CRISTOPHERSEN and PRECHT (1952). He did find a greater amount of water held when dried at specific relative humidities in the case of the cultures grown in the presence of sorbitol. But this was true only if the cultures were not prewashed before drying. In other words, the greater drought tolerance of baker's yeast after growth in the presence of sorbitol followed by washing before exposure to drought, was not associated with any significant increase in the amount of water held by the cells against specific dehydrating forces. Therefore, the drought tolerance of baker's yeast and *Chlorella* cannot be explained by an increased hydration of the protoplasm. FÜCHTBAUER also argues that the positive results of TODD and LEVITT (1951) and of CHRISTOPHERSEN and PRECHT (1952) fail to prove this relation, since the increased water retention may be due to an increase in osmotically effective substances. His objections, however, are invalid. The main reason for the development of the method used by TODD and LEVITT was precisely to eliminate the binding of water by osmotically active substances. For this reason, they first dried the organism at 0% relative humidity until equilibrium was reached. It is obvious that no water can be held osmotically after this treatment. The only water remaining in the tissues is held so firmly as to have essentially no vapor pressure. Calculations showed that

Table 18. *Percentage of survival of baker's yeast after 3 days' exposure to given relative humidities.*
(From FÜCHTBAUER 1957 a.)

Rel. hum. %	Control culture		Control culture 1.1 M sorbitol added before exposure to rel. hum.		Grown in culture plus sorbitol	
	1	2	1	2	1	2
60	41	32	49	36	94	91
40	0	4	5	4	48	63
20	0	0	1	0	24	35
0	0	0	0	0	4	15

Table 19. *Percentage of survival of baker's yeast. Sorbitol cultures washed 3 times with distilled water, then suspended in distilled water before exposure to given relative humidities.*
(From FÜCHTBAUER 1957 a.)

Rel. hum. %	Control Cultures			Sorbitol Cultures		
	1	2	3	1	2	3
90	100	100	100	100	100	100
80	84	80	85	96	100	97
70	87	79	72	99	100	95
60	54	39	31	80	63	73

even the water of hydration of glucose (one molecule per molecule glucose) was far too small to account for the differences obtained.

In spite of his own negative results, FÜCHTBAUER is forced to argue for an increased cytoplasmic hydration from indirect evidence, e. g. the greater cytoplasmic volume in yeast cells grown in hypertonic sorbitol as shown by the unchanged cell size and the smaller vacuoles. But it is difficult to reconcile such a conclusion with his experimental results that fail to show a difference in "bound water." These may, however, be at fault—either because he used relative humidities well above 0% or because he dried his samples only for hours, whereas TODD and LEVITT found that equilibrium could not be attained in less than several weeks. FÜCHTBAUER points to one unfortunate feature of unicellular organisms. The adaptation may not be a true one due to a selection of mutations.

Unlike avoidance, an understanding of drought tolerance requires some knowledge of the nature of drought injury. This injury, however, may not always be of the same type. HENRICI (1946), for instance, gives evidence of what may possibly be an indirect type of injury. Though alfalfa wilted at 3% water loss, Karoo-bush did not wilt even at 30% loss, though the leaves became brittle. This may perhaps hark back to the old idea (MAXIMOV 1929) of a mechanical damage due to bending of wilted leaves. Such damage might be prevented by possession of sufficient strengthening tissue. But this kind of injury could only be local, and the main drought injury is best explained by ILJIN's theory (see LEVITT 1956).

Theories of drought tolerance therefore use ILJIN's concept of drought injury as a basis. STOCKER, for instance, has developed a concept that has been considered elsewhere (LEVITT 1951, 1956). He (1956) has since restated his views with the aid of new evidence, so they must again be analyzed. STOCKER not only fails to recognize the fundamental difference between drought avoidance and tolerance, but he even refuses to distinguish between the effects of drought on growth and on survival. He therefore, attempts to explain all of these phenomena on a single basis. The only evidence he gives for a difference in drought resistance is field experience, i. e. yield differences under conditions of drought. He produces evidence of a greater sensitivity of the stomata in the more resistant oat variety—they open earlier and close more readily at noon. This, he concludes, favors assimilation and reduces water loss. Differences in leaf water content, at least at relative humidities above 50% upheld this concept. Consequently, he has some evidence of a difference in drought avoidance between the two varieties, but no evidence whatever of a difference in drought tolerance. Nevertheless, he considers the differences between these two varieties under the heading of "protoplasmic drought resistance." This leads to difficulties in discussing his theory. However, there seems little doubt that his "protoplasmic drought resistance" is intended to explain drought tolerance, for he states that drought killing is merely the last stage of the drought injury that in smaller doses leads to "protoplasmic drought resistance."

His theory is based first on the assumption of an initial disruption of protoplasmic structure during a brief "reaction phase," and a second teleological assumption that the plant must repair the "breaks" during a longer lasting "restitution phase." He now adds a third assumption based on a Russian theory of enzyme activity. According to this concept, enzymes are free or bound with varying stability on a lipid-protein complex. A loosening of the bonds (e. g. in the "reaction phase") frees the enzymes and increases hydrolytic activity, a tightening of the bonds (e. g., in the "restitution phase") increases synthetic activity.

As evidence for his assumptions, he describes the results of protoplasmic viscosity measurements on *Lamium maculatum* during and following a 24-hour drop in soil water from 100% to 22% water capacity. During the 24-hour drop, the viscosity decreased to half, then rose during the next 11 days to three and a half times the original value.

It should be mentioned that since he is using these results to prove a change in structure, he cannot be dealing with true viscosity, which is a property of fluids. He must mean "structural viscosity" or consistency (SEIFRIZ 1955). One of the most astonishing aspects of the above experiments is the fact that not only was there a difference between the moisture supply in his dry and moist cultures, but also in temperature. His dry cultures were at 36° C., compared to 25° C. for his moist cultures. His results can obviously, therefore, be more easily explained by this temperature difference. Temperatures above 30° C. are known to result in an initial decreased protoplasmic consistency (SEIFRIZ 1955). This could easily be followed by a gradual syneresis, with a transfer of the excess liquid to the vacuole. As a result, the consistency of the protoplasm would gradually rise. The mere fact that the osmotic value rose only during the first 24 hours, then failed to change, showed that no drought hardening was occurring, for other workers have clearly shown that the osmotic value does rise when the plant hardens (LEVITT 1956). In all these experiments, STOCKER and his coworkers tacitly assume, but make no attempt to prove, that the dry cultures become more drought resistant, and that this assumed increase is in drought tolerance as well as avoidance.

The reason for the use of the high temperature for the dry cultures is no doubt the fact mentioned by STOCKER that no "reaction phase" can be detected by such measurements of protoplasmic consistency if the drying is at the normal rate. This also indicates that it is an artificial and abnormal phenomenon. Further proof is the fact that frost hardening by transfer to 5° C. results in an immediate increase in both osmotic potential and protoplasmic permeability that can be detected within 24 hrs. (the time for STOCKER's "reaction phase") and these changes both continue in the same direction for at least several days (LEVITT and SCARTH 1936). There is certainly no oppositely directed "reaction phase" such as STOCKER proposes. Since frost and drought hardiness are correlated (LEVITT 1956), the same must apply to the latter. Further evidence against the above viscosity relationship is provided by ABEL (1956 b). Protoplasmic viscosity was lower in the more drought resistant leaf tip of mosses than in the less resistant cells at the leaf base.

STOCKER also attempts to explain the changes in assimilation and respiration rates during drought on the basis of his two phases and the Russian theory of enzyme action. According to his concept, the "reaction phase" is characterized by hydrolysis or breakdown reactions and therefore respiration increases; the "restitution phase" is characterized by synthesis, and therefore assimilation increases. Such a generalization is obviously invalid, since in opposition to the Russian theory, some enzymes are always firmly attached to protoplasmic structures, others are always free in solution (BONNER 1950). Direct evidence against the generalization is the fact that during frost hardening (and therefore drought hardening since the two increase together) which is STOCKER's "restitution phase," hydrolysis of starch to sugar occurs in dicotyledons, but synthesis of sugars occurs in monocotyledons such as grains (see LEVITT 1956).

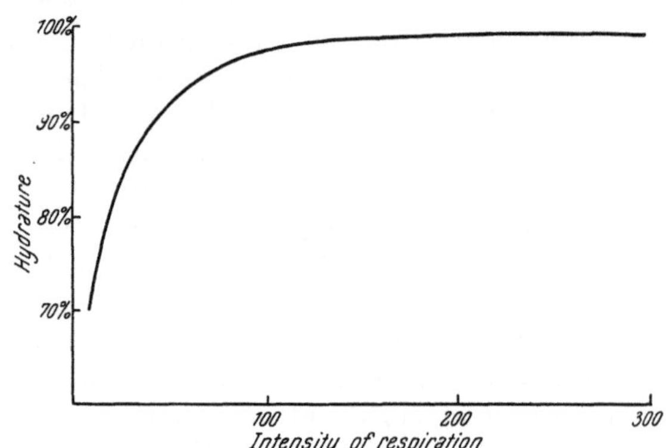

Fig. 21. Relation between respiration rate and hydrature (in % relative humidity) in mosses. (From WALTER 1950.)

Even in the case of drought itself, STOCKER's generalization is not valid. WALTER (1950) found a direct relation between respiration rate and hydration in the case of poikilohydric plants such as mosses (Fig. 21). ILJIN has shown the same relation for higher plants that withstand wilting (Table 20). It is only those plants that are irreversibly injured by decreased water content that show the rise in respiration that STOCKER associates generally with wilting, e. g. hydrophytes (Fig. 22). RAHEJA (1951), in fact, found that

Table 20. *Respiratory intensity of Centaurea scabiosa at various degrees of water loss.*

(From ILJIN—see WALTER 1950.)

Water loss (%)	Relative CO_2 evolution
0	100
29	99
41	84
25	85

drought resistant sugar cane varieties have a lower respiration rate than less resistant varieties. His generalizations are also invalid for assimilation. As in the case of respiration, hydrophytes that are irreversibly injured by dehydration again show the change (a decrease) that he describes for the "restitution phase" of drought resistant plants (Fig. 22).

But, SIMONIS (1947, 1952) clearly demonstrated a greater assimilation in "dry" grown than in "moist" grown plants. This would agree with STOCKER's prediction for his "restitution phase." Unfortunately, SIMONIS did not measure drought resistance in general or tolerance in particular. However, at least three of the four species he investigated (*Andromeda polifolia, Vicia Faba,* and *Nasturtium officinale* R. Br.) are typical mesophytes or even hydrophytes and therefore cannot be expected to develop drought tolerance in "dry" culture. He was also able to detect an increased protoplasmatic viscosity in dry culture, in agreement with STOCKER. All these results indicate that the changes associated with drought described by STOCKER are characteristic of those plants that are unable to develop drought tolerance, and therefore, have no role in the mechanism of drought tolerance.

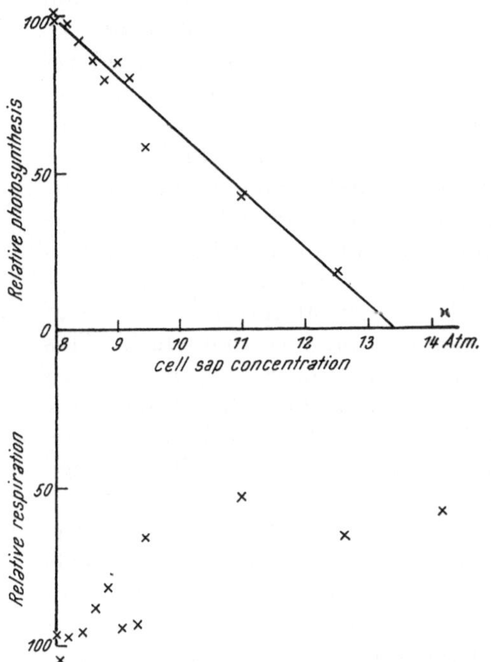

Fig. 22. Relation of CO_2 assimilation and respiration (relative values) in *Elodea canadensis* and hydrature (i. e. cell sap concentration in atms.). Growth stops at 8—10 atms., depending on the temperature.
(From WALTER 1950.)

In agreement with this conclusion are the results of other investigators. PETRIE and ARTHUR (1943) obtained assimilation results with tobacco directly opposed to those of SIMONIS. ECKHARDT (1953) found relatively little change in photosynthesis with change in water balance. The latter investigation was with a mesomorphic xerophyte, i. e. a plant that can be expected to develop drought tolerance under conditions of low water supply.

In his earlier papers, STOCKER attempted to explain the assumed protoplasmic changes during drought by an increased Ca content and a reduced K/Ca ratio, on the basis of some total ash analyses. Since however, a more thorough investigation (WERK 1954) conclusively proved that no such change occurs, STOCKER (1956) now argues that total ash does not reveal the concentration within the protoplasm. FÜCHTBAUER (1957 b), however, has disproved even this argument. He has shown that yeast cells can have their drought tolerance markedly increased by growth in the presence of sorbitol. This treatment failed to change the N, K, or Ca content per unit dry matter. In view of the very small vacuoles in these cells, this result must apply to the protoplasm.

Another complication introduced by STOCKER is an "activation" or a "reactivation" phase which involves the same tendencies as the "restitution phase" but differs by occurring at optimal saturation following too sudden a drying to permit the "restitution phase." A logical conclusion from this would be that a sudden, severe drought followed by rehydration can lead to hardening just as well as a gradual drought. This, of course, is directly opposed to experience.

On the basis of his theory, without a "reaction phase" there would be no "restitution" or hardening phase. In other words, plants never exposed to an injurious environmental factor would possess no resistance. Yet many plants never exposed to drought do possess a remarkable degree of drought resistance, and this is true also of tolerance, which is a protoplasmic character. Good examples of this are given by some woodland and seashore plants (Fig. 1).

In short, STOCKER's theory is basically teleological; the increased respiration rate on drying, for instance, is supposed to supply the greater energy "needed" to repair protoplasmic breaks. It also disagrees completely with the facts. Neither does it take into account the fact that drought resistance is so varied that there is no single common denominator such as he tries to suggest. The most logical explanation of his results is as follows. All the plants he used were apparently devoid of drought tolerance. On subjection to sudden slight drought, they were therefore slightly injured. They gradually recovered from this injury at the same degree of drought due to slow development of drought avoidance characters. This phenomenon has been observed in soybeans (CLEMENTS 1937 a). When drought sets in they wilt during the day and recover at night. After a few days, though the drought is maintained, they no longer wilt during the day. CLARK and LEVITT (1956) showed that this recovery or "hardening" was not accompanied by an increase in drought tolerance. It was completely accounted for by an increased deposit of surface lipids on the leaves which decreased the cuticular transpiration. In this way their water balance became normal again and they no longer wilted. Only when plants are used that increase their drought tolerance on exposure to moderate drought can the associated changes be expected to throw any light on the mechanism of drought tolerance.

HÖFLER's school has raised two arguments against ILJIN's mechanical concept of drought injury, on the basis of results with mosses and liverworts (ABEL 1956 a and b). (1) Though *Fontinalis antipyretica* shows a relatively narrow zone between the "vital boundary" in which all the cells are still alive (at 96% rel. hum.) and the "lethal boundary" in which all the cells are dead (at 90% rel. hum.) other species show a much wider zone. (2) Predrying of *Mnium Seligeri* in 96% rel. hum. lowers the threshold of injury from 72% to 48% relative humidity. Since the cell wall's elasticity must remain essentially constant in both cases, HÖFLER feels that the variations in drought tolerance must be due to a property of the protoplasm and cannot be explained on the basis of ILJIN's concept. This argument, however, is not convincing. When a stress is exerted on a body, the effect will depend both on the magnitude of the stress and on the resistance of the body to the stress. Thus, mechanical injury to protoplasm can be prevented both by reducing the stress exerted by the strained wall and by increasing the elasticity of the protoplasm. Consequently, their results fit in perfectly with ILJIN's mechanical concept of injury.

7. Heat Resistance

Research on heat resistance still lags far behind that on frost and drought resistance. Consequently, problems long since settled for the latter two (see LEVITT 1956) are still being investigated for heat resistance. The question whether or not heat injury occurs in nature continues to be asked. Some have attempted to answer it by comparing plant temperatures in nature and heat killing temperatures in the laboratory. LANGE (1953) showed that lichen temperatures may be well above that of the surrounding atmosphere (Fig. 23) and as high as 69,6° C. under natural conditions. He suggested that they may rise to 75–80° C., since these values have been observed for surface temperatures of soils that are their natural habitat. Since half-hour periods at 70–100° C. were sufficient to kill a large number of different kinds of lichens in the dry state, and since the high temperatures he observed were maintained for much longer than half an hour (Table 21), he concluded that they must be sometimes injured by heat under natural conditions. Both lichens and mosses were much more sensitive to heat in the moist condition (LANGE 1953, 1955). In this state they were killed by temperatures as low as 35–46° C. This fact increases the likelihood of heat injury under natural conditions.

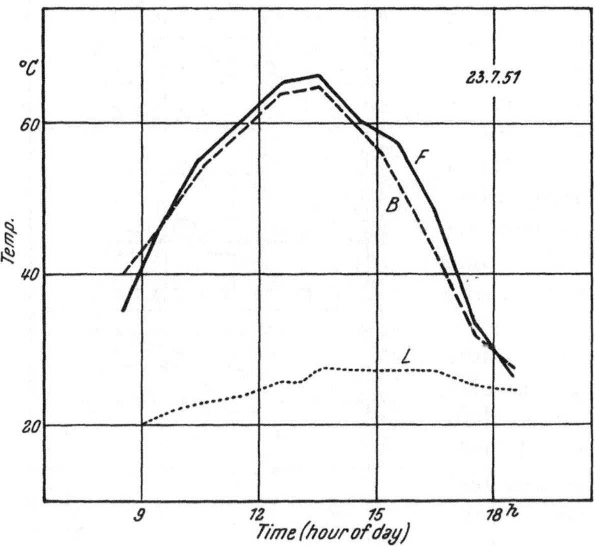

Fig. 23. Diurnal temperature course in the lichen *Cladonia furcata* var. *palomoea (F)*, the soil *(B)*, and the air *(L)*. (From LANGE 1953.)

Table 21. *Time (mins.) during which the temperature of the lichen remained above the listed temperatures (° C.).* (From LANGE 1953.)

Species	62.5°	65.0°	67.5°	Max. temp.
Cladonia pyxidata	270	225	150	69.6°
Cladonia subrangiformis	195	125	—	67.0°
Lecidia decipiens	50	—	—	64.3°

In the case of higher plants, complications enter because they are homoiohydric as contrasted to the poikilohydric lichens and mosses (LANGE 1953). As a result, they are capable of lowering their temperature somewhat by transpiration (see below). Also, due to their greater bulk, they are less likely to reach as high temperatures as the small lower plants. Yet it has long been known that the temperature of higher plants can rise well above that of the surrounding air (LEVITT 1956). This is true, not only of leaves with their large specific surface and high absorption of radiant energy, but also of the stem. PRIMAULT (1954) measured the cambium temperature of fruit trees and found it to be 26° C. (the same as that of the air) on the northeast side, but 37° C. on the southwest side. He even suggested that the "frost furrows" on the surface of trees after a cold spring are really due to overheating of the stems which are at this time unprotected from the sun's rays by the leaves.

This does not agree with the interpretation of FERKL (1951) who was also able to detect fluctuations of more than 20° C. in cambium temperatures of fruit trees on sunny spring days. He ascribed the lesions produced, to frost injury resulting from the temperature differences on the two sides of the trunk. It does not seem possible that the tissue temperature can rise to the killing point during the relatively cool spring. It is far more likely that the dehardening due to this warming results in frost injury during subsequent light frosts.

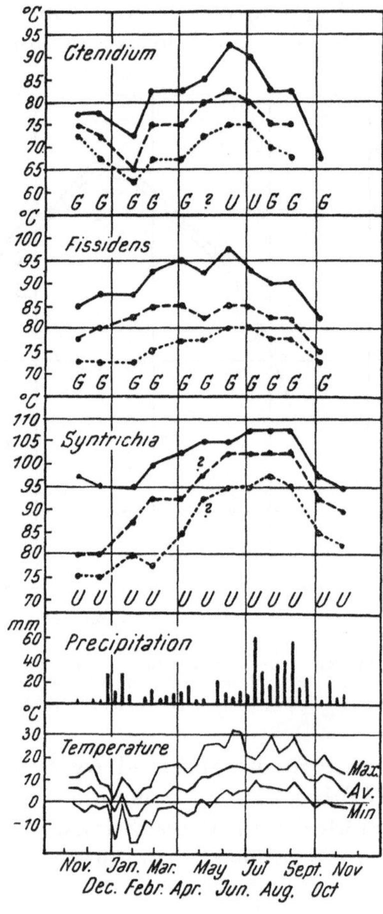

Fig. 24. Seasonal change in heat resistance of mosses exposed in the dry state for a half hour to the given temperatures. Complete killing (solid line), severe injury (broken line), and slight injury (dotted line). G = injured by 50 hour exposure over P_2O_5; U = uninjured. (From LANGE 1955.)

Nevertheless, there is definite evidence of heat injury to higher plants. SZIRMAI (1938) was able to produce the "drought fleck disease" of paprika (which damages 4–12% of the crop in Hungary) artificially by exposures to temperatures of 50–52° C. when the surface was wet. In the case of dry fruit, 55° C. was required to produce the injury. Direct exposure to sunlight produced the symptoms at air temperatures of 49° C.

Little information has heretofore existed as to the variations in heat resistance under natural conditions. LAUDE and CHAUGULE (1953) found the greatest heat resistance in bromegrasses immediately after emergence, even in seedlings deprived of light. Older seedlings deprived of light showed greatly reduced resistance. Resistance diminished in seedlings about 14 days after planting and increased with age after 28 days. In agreement with earlier results on higher plants (SAPPER 1935), LANGE (1953,

1955) found that the heat resistance of lichens and mosses paralleled their habitat. But there were exceptions. The shade lichen *Umbilicaria deusta* and the hydrophytic *Dermatocarpon aquaticum,* both possessed unexpectedly high heat resistance. There were clearcut seasonal changes in the heat resistance of mosses (Fig. 24), with the maximum during summer, the minimum during winter. *Hypnum cupressiforme* showed a very pronounced change in heat resistance with habitat; the heat killing temperature varied from 65–95° C.

An interesting discovery of Lange's is the fact that lichens and mosses differ in heat resistance not only when moist, but also when in the much more resistant dry state—after drying over P_2O_5 to a moisture content of only 0.7–0.8%. These results are somewhat unexpected, in view of the apparent ability of *all* protoplasm to survive the *lowest* temperatures unimpaired when in the dry state (Levitt 1956).

As in the case of drought and frost resistance, heat resistance refers to survival, i. e. the ability of the protoplasm to stay alive at high temperatures. In analogy with xerophily, the term thermophily refers specifically to the ability of an organism to grow at a higher temperature. The two may, of course, go hand in hand though not necessarily controlled by the same factors. A good example of a factor that may result in thermophily without affecting heat resistance is adaptive enzyme formation. Some strains of *Neurospora crassa,* a tropical or subtropical species, are able to produce the enzyme cellulase, apparently adaptively, in the presence of cellulose (Hirsch 1954). But though this occurs at 35° C., it does not occur at 25° C. The organism may therefore be considered thermophilic on cellulose, a factor that could not possibly affect its heat resistance.

The main methods for measuring heat resistance are still the standard ones previously described (Levitt 1956). Laude and Chaugule (1953) continue to use a method that fails to differentiate categorically between heat and drought resistance. They exposed plants to a temperature of 130° F. and a relative humidity of 30–35%. Though the exposure periods were short (4¾ hrs.), it still remains to be proved that drought injury does not occur at these low relative humidities and high temperatures.

Lange (1953) measured the heat resistance of lichens by determining the temperature that reduces subsequent equilibrium respiration to 50% of normal. Actual growth of the lichens gave similar results. This indicated that the lichen fungus, which is primarily responsible for the respiration, has about the same heat resistance as the lichen alga, which is mainly responsible for the resumption of growth.

Avoidance. The ability of a plant to avoid heat injury would be indicated if it could maintain its temperature below that of its environment. Leaf temperatures, on the contrary, are higher than air temperatures during hot periods (Levitt 1956). The same is also true of cambium temperatures when the stem is exposed to direct radiation from the sun (see above). However, those plants that show the smallest rise in temperature would have to be accepted as heat avoiding. This could conceivably be accomplished in several different ways.

1. Insulation. It does not seem likely that plants could possess heat avoidance due to insulation from the surrounding environment. On the contrary, this would help them to store up the absorbed radiant energy and to reach still higher temperatures than if they could conduct the heat away (e. g. the cambium temperatures mentioned above). However, there is one case where insulation may be important. If injury occurs at the soil level due to excessive heating of the surface soil, the cambium could be protected from heat injury by the insulation of bark. Indirect evidence of this is the fact that such injury occurs only to young seedlings that have little such protection (see LEVITT 1956). But other explanations of this fact are just as reasonable.

2. Heat of vaporization. The loss of water from a plant is, of course, accompanied by a loss of heat due to the heat of vaporization of water. The importance of this factor has been carefully analyzed by CURTIS and is reviewed by CURTIS and CLARK (1950). According to their calculations, the heat absorbed by the water lost from a rapidly transpiring leaf at 40–50° C. could account for only about 15% of the radiant energy absorbed in full sunlight. In exceptional cases, however, they admit that the transpiration rate is high enough to have a greater effect, though seldom if ever sufficient to maintain the temperature of a leaf in direct sunlight as low as that of the surrounding air.

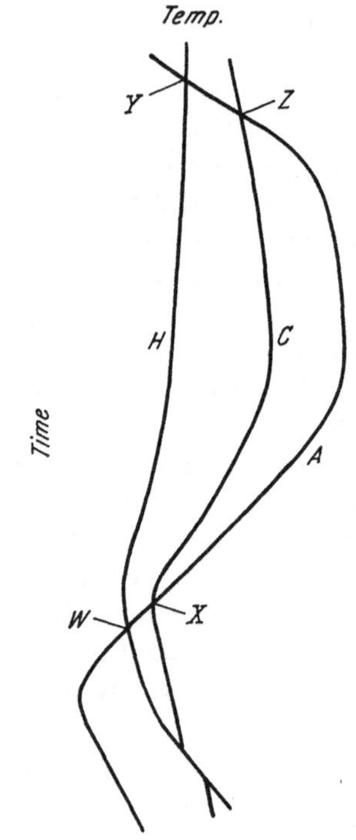

Many measurements of leaf temperatures have indicated large decreases (e. g. 20–30° C.) due to transpiration. From CURTIS' critical analysis of these results, he concluded that the maximum temperature drop due to transpiration is only about 5 to 6° C. Even on the basis of this conservative estimate, there may be many cases when the transpiration would mean the difference between reaching the killing temperature and going no higher than a non-injurious temperature of 5 or 6° below that point.

Fig. 25. Temperature curves for air (A), tree center (H), and cambium (C) for a 24-hour period from 7 a. m. W and X are a. m., Y and Z are p. m. intersections with air temperature. Temperature increases from left to right, time from bottom to top. (From REYNOLDS 1939.)

Another possible role of the heat of vaporization of water has received little attention. According to REYNOLDS (1939), due to the tension in the sapwood of rapidly transpiring plants, the vapor pressure of the ascending sap decreases to the point where distillation occurs into it from the heartwood. These is, therefore, a temperature gradient from a low in the heartwood to a high in the surrounding air, the cambium temperature lying somewhere between the two (Fig. 25). Under such conditions, overheating of the cambium could not occur. Presumably the heat release on condensation of the water in the sapwood is carried away by the ascending stream.

3. Reduced respiration. Though this may actually occur, it cannot be a factor of any importance since the amount of energy released is only a minute fraction of the radiant energy absorbed or even of the heat of vaporization of the water transpired. In rapidly growing organs (e. g. *Arum* spadix—see Levitt 1956) a marked temperature rise may occur due to respiration. But such growth normally takes place at a time of the year when the air temperatures are not high.

4. Reduced absorption of radiant energy. Since the rise in plant temperatures above that of the surrounding air is due to the greater absorption of radiant energy from the sun, it is logical to look for heat avoidance due to a reduced absorption of radiant energy. This may conceivably be due to an increase in either the reflectance or transmissivity of the plant tissues.

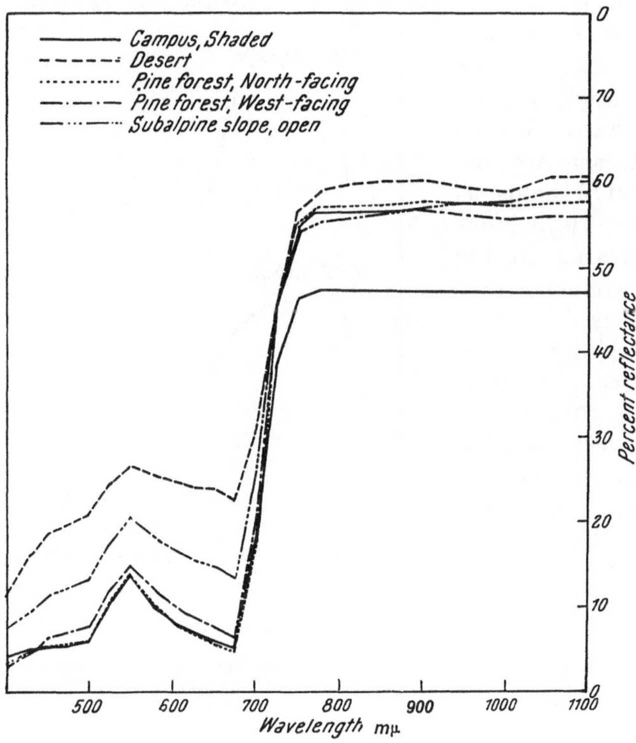

Fig. 26. Percentage reflectance from leaves of various species. (From Billings and Morris 1951.)

a) Reflectance. Some evidence of a greater reflectance by species of hot habitats has been produced by Billings and Morris (1951). In general, the reflectance curves of green leaves showed values of about 5% at 440 mμ, rising to a peak of about 15% at 550 mμ, with a gradual slope down to 5 or 6% at 675 mμ. Above this, the curves rose steeply to a plateau of about 50% in the infrared region of 775–1100 mμ. The hairs or scales on desert and subalpine plants were correlated with higher reflectance in the visible but not necessarily in the infrared region. On the average, desert species reflected the greatest amount of visible radiation, followed by subalpine, west-facing pine forest, north-facing pine forest, and shaded campus species, in that order (Fig. 26). In the infrared, the differences between groups were not so marked, but the greatest reflectance here also was shown by the desert species, with an average value of about 60%. The greatest infrared reflectance (almost 70%), was measured from the glabrous leaves of the desert peach, *Prunus andersonii.*

On a clear day, considerably more than 50% (often as much as 65%) of the solar radiation incident at the earth's surface is in the infrared (GATES and TANTRAPORN 1952). The major effect on the plant's temperature must therefore be produced by reflectance in the infrared rather than in the visible, the more so since the plant reflects so little of the latter. On the basis of BILLINGS and MORRIS' results, shade leaves absorb about 1/3 more in the infrared than do desert plants. Their temperatures might therefore be expected to rise proportionately higher.

But this assumes that the reflectance found by BILLINGS and MORRIS for the short infrared (up to 1100 μ) holds as well for the long infrared which includes most of the incident infrared in the radiation from the sun. GATES and TANTRAPORN (1952) have measured the reflectance by leaves in this long infrared region (1–20 μ). They find that it is generally small—less than 10% for an angle of incidence of 65°, less than 5% for 20°. Since the transmissivity of the leaves is zero in the infrared region beyond 1.0 μ, this means that the absorption varies from above 90% to below 100% of the incident infrared. Such small differences can have little effect on leaf temperatures. Furthermore, *Opuntia* leaves showed nearly the lowest reflectivity of the 27 species tested, and the hairy leaves of *Verbascum thapsus* failed to reflect any. On the other hand, *Citrus limonia* reflected the most. Shade leaves, in fact, reflected more than sun leaves of the same plant.

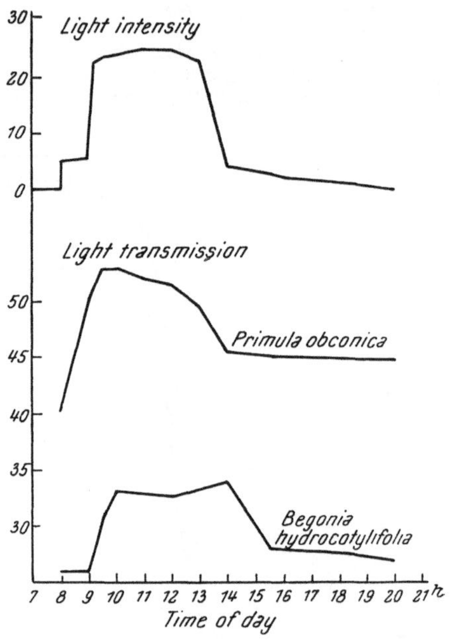

Fig. 27. Diurnal increase of transmissivity by leaves, with increase in light intensity. (From BIEBL 1955.)

Therefore, in spite of the promising differences found in the short infrared, the results with the long infrared (including most of the infrared radiation), lead to the conclusion that plants have not succeeded in developing heat avoidance by increasing their reflectance of the incident radiation.

b) Transmissivity. BIEBL (1955) observed a parallel between the light transmission curves of leaves of *Begonia tuberosa* growing in the open, and the intensity incident on the plant. This was due to the plastid orientation which is very light sensitive. *B. hydrocotylifolia* and *Primula obconica* give similar results (Fig. 27). On the other hand, many native plants growing in the open failed to show this relation. On the contrary. the light transmission often decreased somewhat at midday when light intensity was highest. BIEBL ascribes this to the reduced water content at midday, partly on the basis of the parallel between the relative humidity curve and the light transmission curve. He also felt that starch synthesis was partly responsible.

The increased light transmission with light intensity was not confined to succulents such as *Begonia;* it was just as pronounced in *Primula obconica.* There is, therefore, on apparent relation to heat resistance. This is even more obvious from the fact mentioned above that most of the incident radiation is in the infrared, *none* of which is transmitted. Thus neither reflectivity nor transmissivity differences between plants seem capable of giving rise to heat avoidance.

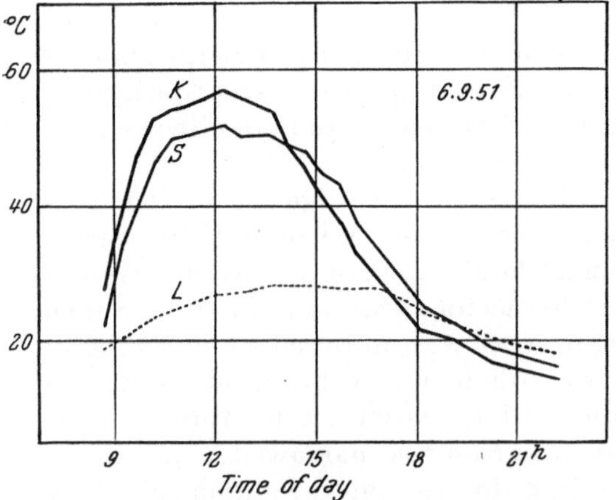

Fig. 28. Diurnal temperature change in a lichen on cork (*K*) and on sandstone (*S*). (*L*) = air temperature. (From LANGE 1954.)

A change in the angle of the leaf blade will, of course, markedly affect the absorption of the sun's rays, and may conceivably lead to heat avoidance. This adaptation is apparently widespread in hot, dry regions (CURTIS and CLARK 1950), though its relation to heat resistance has never been proved.

In the case of mosses and lichens, no avoidance seems to exist. The temperature of the tissues is most markedly affected by the substrate (Fig. 28), and the height above ground (LANGE 1954). During insolation, the temperature may markedly exceed that of the substrate. Their main protection appears to be associated with their extreme drought tolerance, for when the temperature rises they lose water readily and are quickly converted to the air dry state in which they are far more heat tolerant than when moist.

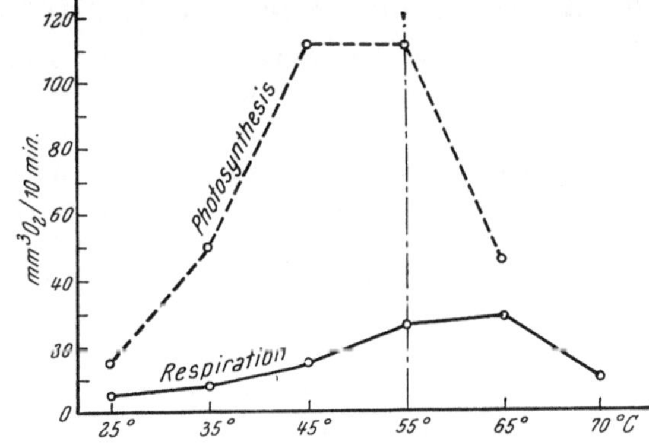

Fig. 29. Change in photosynthesis and respiration with temperature. Alga acclimatized at 55° C. (From MARRÉ and SERVETTAZ 1956.)

From all this evidence heat avoidance seems to play a small, though at times perhaps decisive role in heat resistance. Insulation is of possible value in the sole case of plant parts in contact with the soil surface, which may reach temperatures above the killing point. Heat of vaporization of water may very definitely save a plant from attaining the heat killing point. Reduced respiration rate cannot play a significant part. Reduced absorption of radiant energy seems unable to produce a significant difference between plants, since 90–100% of the longer infrared is absorbed by all plants, and this accounts for by far the greater part of the energy absorbed. The most important factor in heat resistance must therefore be heat tolerance.

Heat tolerance of oats has recently been investigated by COFFMAN (1957) using 45 min. exposures to temperatures of 48.5–51° C. The genetic evidence indicated that three factors were involved and that heat susceptibility was dominant. Tolerance was greater in the early boot stage than either earlier or later.

MARRÉ (1956, 1957) has recently initiated an investigation of the basis of heat tolerance in thermophytic blue-green algae. When he (1956) varied the acclimatization temperature from 22 to 65° C., both photosynthesis and respiration rose to a maximum near the acclimatization temperature then fell off (Fig. 29). The inactivation process was initiated at a distinctly lower temperature for photosynthesis than for respiration, and was irreversible. The margin between the acclimatization temperature and the irreversible inactivation temperature progressively narrowed with rise in acclimatization temperature. According to his interpretation of these results, irreversible denaturation of proteins and lipoproteins is the main factor in heat resistance. In favor of this interpretation, extracts of TPNH—cytochrome C reductase from a hotspring blue-green alga (*Aphanocapsa thermalis*) cultured at 55° C. were more resistant to heat inactivation than similar extracts from a blue-green alga (*Anabaena cylindrica*) cultured at usual temperatures. He concluded that the heat resistance of thermal algae was due to the structure of the protein molecules.

8. Relation between Frost, Drought, and Heat Tolerance

Since frost and heat resistance are primarily due to tolerance, whereas drought resistance is at least as often due to avoidance, one must be cautious in attempting to find a relation between the three. It is obvious that a drought resistant plant possessing avoidance but no tolerance cannot be expected to show frost or heat resistance. When comparing drought resistance with the other two, it is therefore, always necessary to determine at the outset which kind of resistance is involved. Only too often, if a plant shows an increase in drought resistance on exposure to moderate drought, this is tacitly assumed to be due to increased tolerance. But such a conclusion does not necessarily hold true. PETINOV and ZAK (1938) showed that so-called drought hardening of wheat results in the development of xeromorphic (i. e. drought avoiding) characters, as has long been known for many plants (MAXIMOV 1929). Whether or not tolerance is also involved was not determined. CLARK and LEVITT (1956) found only a pseudo-hardening in soybeans; they developed increased avoidance but no increased tolerance on exposure to moderate drought. Other factors besides drought may also alter these xeromorphic characters, e. g. N-fertilization, which decreases the xeromorphy of high moor plants (MUELLER-STOLL 1947).

Another source of error in comparing the three kinds of resistance is the method of measurement used. OCHI (1952 b) was unable to find a clear cut relation between frost and drought resistance; but he determined the latter by the drying time required to kill the plant. No attempt was made to determine the degree of dehydration, or even to control the external environment.

In any attempt to determine whether or not there is a relation between the three kinds of resistance, the following steps are therefore necessary:

1. A quantitative determination under standard conditions of the number of degrees of frost, the percent relative humidity deficit, or the number of degrees of heat respectively, that kills about 50% of the plant.

Table 22. *Comparison of absolute values for frost, drought, and osmotic tolerance in Enteromorpha clathrata (Roth) Greville.* Freezing point of seawater taken as — 2.0 to — 2.3° C.
(From BIEBL 1956.)

	Freezing for 21 hrs.	Desiccation for 14 hrs.	Osmotic concentration for 24 hrs.
Maximum survived without injury	— 15° C.	83.9% r. h.	5.0 × seawater
Tolerance	14° C.	15%	9 to 10.5° C. (app.)

2. Control of the conditions to assure that tolerance and not avoidance is measured. This can easily be done by (a) inoculating the plant with an ice crystal when it is just below its freezing point; (b) by exposing unprotected sections of the plant to atmospheres of known relative humidities; or (c) by plunging sections of the tissues into water at the known temperatures, or in the case of whole plants, exposing them to known temperatures in a saturated atmosphere.

Only if such tests prove that tolerance is the cause of resistance, can any relation between the three kinds of resistance be expected. All the investigations that have taken these precautions have overwhelmingly indicated a clearcut parallel between frost, drought, and heat tolerance or hardiness (LEVITT 1956). It has also been shown (LEVITT 1956) that the few available absolute measurements of frost and drought tolerance in the same plant give identical results within the limits of error of the method used. BIEBL (1956) has recently provided further results that can also be recalculated on a comparative basis. The frost hardiness and drought hardiness agree within the limits of error of the method used (Table 22). The calculated osmotic hardiness is a little low, but this is no doubt at least partly due to the fact that concentrating seawater 5× must lower the freezing point more than the calculated 5×. It is also possible that such high concentrations of salts may be toxic. BIEBL also determined the chemical resistance of the plants, but this was not related to the other three kinds of tolerance. Since less evidence is available for a relation between frost and heat tolerance than between the other two combinations,

COFFMAN's recent evidence is welcome (Table 23). But it might be objected that this parallelism does not prove that they all depend on the same protective factors in the plant. One might, for instance, logically expect the development of heat resistance in plants whose drought resistance is due to a conservation of water. These plants are unable to resort to the cooling effect of the heat of vaporization of water; consequently, their

Table 23. *Relative winter hardiness and heat resistance of 12 varieties of oats (Fulghum check, 100 percent).* (From COFFMAN 1957.)

Type and variety	*Avena* species	Winter survival (%)	Heat survival (%)
	True winter		
Fulwin	*A. byzantina*	137.6	123.4
Bicknell	*A. sativa*	130.1	124.7
Hairy Culberson	*A. sativa*	129.8	122.2
Pentagon	*A. byzantina*	128.2	103.1[1]
Tech (V. P. I. No. 1)	*A. sativa*	125.1	105.5[1]
Winter Turf	*A. sativa*	120.1	111.2
Lee	*A. sativa*	116.8	100.0[1]
Culred	*A. byzantina*	108.5	120.0
	Intermediate winter		
Appler	*A. byzantina*	100.3	99.8
Fulghum (check)	*A. byzantina*	100.0	100.0
	Spring		
Bond	*A. byzantina*	63.8	68.2
Victoria	*A. byzantina*	46.8	43.0

[1] Additional data, not fully comparable with those shown, reveal that this variety is somewhat more heat resistant than indicated here.

temperature will rise somewhat above that of plants that do. In order to survive, they must, therefore, have developed heat resistance as well as drought resistance. But such a parallelism is very different from that between drought tolerance and heat tolerance. In the former case the parallelism is fortuitous, due to the simultaneous selective actions of heat and drought; in the latter it is inevitable due to the dependance of both kinds of tolerance on the same internal plant factors.

The evidence in both the older and the newer literature (LEVITT 1956) clearly shows that the relation is not simply fortuitous. Some recent results by LANGE (1953, 1955) are particularly appropriate. He noted a high resistance in lichens and mosses to both heat and drought in dry, hot regions; a low resistance in cool, moist regions. As mentioned above, lichens and mosses are incapable of drought avoidance since they are poikilohydric. Consequently, this is unquestionably a case of tolerance. The only question is whether the parallelism is fortuitous or due to the same internal factors. His observations indicated that the plants are never exposed to a drought capable of producing injury, but that some of them may be exposed to temperatures that go above their maximum.

The most clearcut relation was found in a few exceptional species. The shade lichen *Umbilicaria deusta* possessed an unexpectedly high drought resistance, and its heat resistance was also high. Since this species does not grow under conditions of excessive drought or heat, the parallelism cannot be explained by independent selection. Similarly, *Dermatocarpon aquaticum* had a high heat resistance, even though it is a hydrophyte. On the other hand, species (e. g. *Lobaria pulmonaria*) sensitive to heat were also drought sensitive. Mosses also showed a parallelism between drought and heat resistance (Lange 1955). *Clenidium* was injured by drying over P_2O_5 for 50 hrs., except during June and July when its heat resistance was at a maximum (Fig. 24). The reduction in heat resistance (tested in the dry state) was also related to the time it was kept in the wet state.

These results of Lange's are particularly welcome since until now there was an overabundant amount of evidence relating frost and drought tolerance, but not as much relating heat and drought tolerance. The relation between the three is now well established, and is in agreement with the theory that the same kinds of protoplasmic changes are associated with the development of all three kinds of tolerance (Levitt 1956). No new theory, nor any valid objection to the theory previously proposed (Levitt 1956) has arisen in the newer literature.

Bibliography

Abel, W. O., 1956 a: Die Austrocknungsresistenz der Laubmoose. S.ber. Öst. Akad. Wiss., math.-naturw. Kl. **10**, 105—107.

— 1956 b: Die Austrocknungsresistenz der Laubmoose. S.ber. Öst. Akad. Wiss., math.-naturw. Kl. **165**, 619—707.

Allen, R. M., 1955: Foliage treatments improve survival of longleaf pine plantings. J. For. **53**, 724—727.

Amirshahi, M. C., and F. L. Patterson, 1956: Development of a standard artificial freezing technique for evaluating cold resistance of oats. Agron. J. **48**, 181—184.

Anderson, L. E., and P. F. Bourdeau, 1955: Water relations in two species of terrestrial mosses. Ecology **36**, 205—212.

Andrew, R. H., 1954: Breeding maize for cold resistance. A review. Euphytica **3**, 108—116.

Aoki, Kiyoshi, 1948: Analysis of the Freezing Process of Living Organisms. III. Two Freezing points on the Freezing Curve of Plant Tissues. Low Temp. Science **4**, 66—77.

— 1950: Analysis of the Freezing Process of living Organisms. I. The Relation between the Shape of the Freezing Curve and the Mode of Freezing in Plant Tissues. Low Temp. Science **3**, 219—227.

— Eizo Asahina, and Isao Terumoto, 1955: Analysis of Freezing Process of Living Organisms. IX. The Relation between the Shape of the Freezing Curve and the Frost Hardiness. Low Temp. Science **10**, 69—79.

Arvidsson, Inga, 1951: Austrocknungs- und Dürreresistenzverhältnisse einiger Repräsentanten öländischer Pflanzenvereine nebst Bemerkungen über Wasserabsorption durch oberirdische Organe. Oikos. Acta Oecologica Scandinavica. Supplementum **1**, 1—181.

ASAHINA, EIZO, 1954: A Process of Injury Induced by the Formation of Spring Frost on Potato Sprout. Low Temp. Science, Ser. B, **11**, 13—21.
— 1956: The Freezing Process of Plant Cell. Contrib. Inst. Low Temp. Science **10**, 83—126.
ASAHINA, K., K. AOKI, and J. SHINOZAKI, 1954: The freezing process of frost-hardy caterpillars. Bull. Entomol. Res. **45**, 329—339.
BAUMAN, L., 1957: Über die Beziehungen zwischen Hydratur und Ertrag. Ber. dtsch. bot. Ges. **70**, 67—78.
BERNSTEIN, L., and G. A. PEARSON, 1954: Influence of integrated moisture stress achieved by varying the osmotic pressure of culture solutions on growth of tomato and pepper plants. Soil Sci. **77**, 355—368.
BIEBL, R. 1952: Ecological and non-environmental constitutional resistance of the protoplasm of marine algae. J. Marine Biol. Assoc. United Kingdom **31**, 307—315.
— 1955: Tagesgänge der Lichttransmission verschiedener Blätter. Flora **142**, 280—294.
— 1956: Zellphysiologisch-ökologische Untersuchungen an *Enteromorpha clathrata* (Rath) Greville. Ber. dtsch. bot. Ges. **69**, 75—86.
BIEL, E. R., A. V. HAVENS, and M. A. SPRAGUE, 1955: Some extreme temperature fluctuation experienced by living plant tissue during winter in New Jersey. Bull. Amer. Meteorol. Soc. **36**, 159—162.
BILLINGS, W. D., and R. J. MORRIS, 1951: Reflection of visible and infrared radiation from leaves of different ecological groups. Amer. J. Bot. **38**, 327—331.
BONNER, J., 1950: Plant Biochemistry. Academic Press. New York.
BREAZEALE, J. F., and W. T. MCGEORGE, 1949: A new method of determining the wilting percentage of soil. Soil Sci. **68**, 371—374.
— — and J. F. BREAZEALE, 1950: Moisture absorption by plants from an atmosphere of high humidity. Plant Physiol. **25**, 413—419.
BRIERLEY, W. G., R. H. LANDON, and R. J. STADTHERR, 1952: The effects of daily alternations between 27 and 39 degrees F. on retention or loss of cold resistance in the Latham raspberry. Proc. Amer. Soc. Hort. Sci. **59**, 173—176.
— — 1954: Effects of dehardening and rehardening treatments upon cold resistance and injury of Latham raspberry canes. Proc. Amer. Soc. Hort. Sci. **63**, 173—178.
BROUWER, R., 1956: Radiation intensity and transpiration. Neth. J. Agr. Sci. **4**, 43—48.
BULA, R. J., and DALE SMITH, 1954: Cold resistance and chemical composition in over-wintering alfalfa, red clover, and sweet clover. Agron. J. **46**, 397—401.
— — and H. J. HODGSON, 1956: Cold resistance in alfalfa at two diverse latitudes. Agron. J. **48**, 153—156.
BURCIK, E., 1950: Über die Beziehungen zwischen Hydratur und Wachstum bei Bakterien und Hefen. Arch. Mikrobiol. **15**, 203—235.
CARRIER, L. E., 1951: A study of methods of determining the extent of frost injury of roses. Proc. Amer. Soc. Hort. Sci. **58**, 350—356.
CHANDLER. W. H., 1954: Cold resistance in horticultural plants; A review. Proc. Amer. Soc. Hort. Sci.**64**, 552—572.
CHESSIN, M., 1953: Frost resistance in cheatgrass. *Bromus tectorum* L. Proc. Montana Acad. Sci. **13**, 43.
CHRISTOPHERSEN. J., und H. PRECHT, 1952: Untersuchungen über die Bedeutung des Wassergehaltes von Hefezellen für Temperaturanpassungen. Arch. Mikrobiol. **18**, 32—48.
— — 1956: Über die Kälteresistenz von Hefezellen. Biol. Zbl. **75**, 612—624.
CLARK, J. A., and J. LEVITT, 1956: The basis of drought resistance in the soybean plant. Physiol. Plant. **9**, 598—606.
CLEMENTS, H. F., 1937 a: Studies in drought resistance of the soybean. Res. Studies State Coll. Wash. **5**, 1—16.
— 1937 b: Studies in the drought resistance of the sunflower and the potato. Res. Studies State Coll. Wash. **5**, 81—98.
COFFMAN, Fr. A., 1955: Results from uniform winter hardiness nurseries of oats for the five years 1947—1951, inclusive. Agron. J. **47**, 54—57.
— 1957: Cold resistant oat varieties also resistant to heat. Science **125**, 1298—1299.
COOPER, W. C., B. S. GORTON, and S. D. TAYLOE, 1954: Freezing tests with small trees and detached leaves of grapefruit. Proc. Amer. Soc. Hort. Sci. **63**, 167—172.
— and A. PEYNADO, 1955: Effects of plant regulators on dormancy, coldhardiness and leaf form of grapefruit trees. Proc. Amer. Soc. Hort. Sci. **66**, 100—110.
CORNS, G., and G. SCHWERDTFEGER, 1954: Improvement in low temperature resistance of sugar beet and garden beet seedlings treated with sodium TCA and Dalapon. Canad. J. Agric. Sci. **34**, 639—641.

CRANE, J. C., 1954: Frost resistance, and reduction in drop of injured apricot fruits effected by 2, 4, 5-trichlorophenoxyacetic acid. Proc. Amer. Soc. Hort. Sci. **64**, 225—231.

CURTIS, O. F., and D. G. CLARK, 1950: An Introduction to Plant Physiology. McGraw-Hill, New York.

DUISBERG, P. C., 1952: Some relationships between xerophytism and the content of resin, nordihydroguaiaretic acid and protein of *Larrea divaricata* Cav. Plant Physiol. **27**, 769—777.

EAKS, L., and L. L. MACHLIS, 1956: Respiration of cucumber fruits associated with physiological injury at chilling temperatures. Plant Physiol. **31**, 308—314.

ECKHARDT, F., 1952: Rapports entre la grandeur des feuilles et le comportement physiologique chez les xerophytes. Physiol. Plantarum **5**, 52—69.

— 1953: Transpiration et photosynthèse chez un xerophyte mésomorphe. Physiol. Plantarum **6**, 253—261.

EDGERTON, L. J., 1954: Fluctuations in the cold hardiness of peach flower buds during rest period and dormancy. Proc. Amer. Soc. Hort. Sci. **64**, 175—180.

— and M. B. HOFFMAN, 1952: The effect of thinning peaches with bloom and post-bloom sprays on the cold hardiness, of the fruit buds. Proc. Amer. Soc. Hort. Sci. **60**, 155—159.

EMMERT, F. H., and S. HOWLETT, 1953: Electrolytic determination of the resistance of fifty-five apple varieties to low temperatures. Proc. Amer. Soc. Hort. Sci. 62, 311—318.

FERKL, F., 1951: The results of temperature measurement in fruit trees. Sbornik Cesk. Akad. Zemedelske **24**, 175—180 (Biological Abstracts **7**, 20431, 1953).

FERRI, Mario G., 1953: Water balance of plants from the "Caatinga." Rev. Brasil. Biol. **13**, 237—244.

— 1955: Contribuição ao Conhecimento da Ecologia do Cerrado e da Caatings. Univers. São Paulo, Fac. Fil. Cien. Let. Bol. **195**, Bot. 12, 170 pp.

FRYXELL, P. A., 1954: A procedure of possible value in plant breeding. Agron. J. **46**, 433—434.

FÜCHTBAUER, W., 1957 a: Trockenresistenzsteigerung nach osmotischer Adaptation bei *Saccharomyces* und *Chlorella*. Arch. Mikrobiol. **26**, 209—230.

— 1957 b: Über den Zusammenhang der Trockenresistenz einiger Einzeller mit ihrem Retentionswasser und Mineralstoffgehalt. Arch. Mikrobiol. **26**, 231—253.

GATES, D. M., and WIROJANA TANTRAPORN, 1952: The reflectivity of deciduous trees and herbaceous plants in the infrared to 25 microns. Science **115**, 613—616.

GESSNER, F., and F. ZWERENZ, 1950: Über die Erhöhung der Kälteresistenz der Pflanzen bei Stickstoffmangel. Naturw. **37**, 453—454.

GREB, H., 1957: Der Einfluß tiefer Temperatur auf die Wasser- und Stickstoffaufnahme der Pflanzen und ihre Bedeutung für das „Xeromorphieproblem". Planta **48**, 523—563.

GUNAR, I. I., and M. N. SILEVA, 1954: Variation of the sugar content of winter wheats during the hardening-off process. Fiziol. Rastenii **1**, 141—145 (Biol. Abst. **31** (6), 18671. 1957).

HAINES, F. M., 1952: The absorption of water by leaves in an atmosphere of high humidity. J. Exper. Bot. **3**, 95—98.

— 1953: The absorption of water by leaves in fogged air. J. Exper. Bot. **4**, 106—107.

HARRISON, A. P. Jr., 1956: Causes of death of bacteria in frozen suspensions. Ant. v. Leeuw. **22**, 407—418.

HELLSTRÖM, N., 1954 a: Investigations on oil turnips and oil rape. I. A survey of composition in connection with hardening. Acta Agric. Scand. **4**, 302—310.

— 1954 b: Investigations on oil turnips and oil rape. II. A survey of carbohydrate content in connection with hardening. Acta Agric. Scand. **4**, 311—316.

— 1955: Investigations on oil turnips and oil rape. III. A survey of starch content in connection with hardening. Acta Agric. Scand. **5**, 23—30.

— 1956: Investigations on oil turnips and oil rape. VI. Further investigations in chemical composition. Acta Agric. Scand. **6**, 17—44.

— and B. TORSSELL, 1955: Investigations on oil turnips and oil rape. V. On the rate of hardening. Acta Agric. Scand. **5**, 39—43.

HELMERICK, R. H., and R. P. PFEIFER, 1954: Differential varietal responses of winter wheat germination and early growth to controlled limited moisture conditions. Agron. J. **46**, 560—562.

HENRICI, M., 1946: Effect of excessive water loss and wilting on the life of plants. Union S. Africa Dept. Agric. and Forest Sci. Bull. **256**, 1—22.

HIRSCH, H. M., 1954: Temperature-dependent cellulose production by *Neurospora crassa* and its ecological implications. Experientia **10**, 180—182.

HODGSON, H. J., and R. J. BULA, 1956: Hardening behavior of sweetclover (*Melilotus* sp.) varieties in a subarctic environment. Agron. J. **48**, 157—160.

HÖFLER, K., 1951: Cold resistance of some upland bog algae. Verh. Zool.-Bot. Ges. Wien **92**, 234—242 (Biol. Abstr. 25, 32786. 1951).

HUBER B., 1931: Die Trockenanpassungen in der Wipfelregion der Bäume und ihre Bedeutung für das Xerophytenproblem. J. Ecol. **19**, 283—291.

JENNY, J., 1953: Trials of measuring winter temperatures of apple and pear buds. Rev. Romande Agric. Viticult. et Arboricult. **9**, 26—27.

JEREMIAS, K., 1956: Zur Physiologie der Frosthärtung (Unter besonderer Berücksichtigung von Winterweizen). Planta **47**, 81—104.

JOHANSSON, N. O., C. E. ALBERTSON, och T. MÅNSSON, 1955: Undersökningar över höstetets härdning och avhärdning. Sver. Utsäds. Tidsk. 82—96.

— and B. TORSSELL, 1956: Field trials with a portable refrigerator. Acta Agric. Scand. **6**, 81—99.

KANWISHER, J. W., 1955: Freezing in intertidal animals. Biol. Bull. (Am.) **109**, 56—63.

KATO, R., 1951: An attempt to the rational elucidation of social plant life; hydrature and *Abies-Picea* climax in Japan. II. Seasonal change of osmotic pressure of cell sap of dominant trees in coniferous forest. J. Japanese Forest Soc. **33**, 215—218.

KILIAN, CH., et G. LEMÉE, 1956: Les Xérophytes: Leur économie d'eau. Handbuch der Pflanzenphysiologie **3**, 787—824.

KOFRANEK, M., 1952: Some effects of low soil temperature upon the growth of *Euphorbia pulcherrima*. Proc. Amer. Soc. Hort. Sci. **59**, 509—515.

KONDO, Y., 1952; Physiological studies on cool-weather resistance of rice varieties. Bull. nat. Inst. Agric. Sci. (Japan), ser. D. Plant Physiol. **3**, 113—228.

KONOVALOV, I. N., 1955: Experiments on increasing the frost resistance of stock and cabbage by the action of extracts of winter-resisting plants. Doklady Akad. Nauk U. S. S. R. **101**, 767—770.

KRAMER, P. J., 1950: Effects of wilting on the subsequent intake of water by plants. Amer. J. Bot. **37**, 280—284.

KURSANOV, H. L., 1956: Recent advances in plant physiology in the U. S. S. R. Ann. Rev. Plant Physiol. **7**, 401—436.

LANGE, O. L. 1953: Hitze und Trockenresistenz der Flechten in Beziehung zu ihrer Verbreitung. Flora **140**, 39—97.

— 1954: Einige Messungen zum Wärmehaushalt poikilohydrer Flechten und Moose. Arch. Meteor. Geoph. u. Bioklim., Ser. B, **5**, 182—190.

— 1955: Untersuchungen über die Hitzeresistenz der Moose in Beziehung zu ihrer Verbreitung. I. Die Resistenz stark ausgetrockneter Moose. Flora **142**, 381—399.

LARCHER, W., 1954: Die Kälteresistenz mediterraner Immergrüner und ihre Beeinflußbarkeit. Planta **44**, 607—635.

LAUDE, H. M., and B. A. CHAUGULE, 1953: Effect of stage of seedling development upon heat tolerance in bromegrasses. J. Range Management **6**, 320—324.

LEVITT, J., 1951: Frost, drought, and heat resistance. Ann. Rev. Plant Physiol. **2**, 245—268.

— 1953: Further remarks on the thermodynamics of active (non-osmotic) water absorption. Physiol. Plant. **6**, 240—252.

— 1954: The role of carbohydrates in frost resistance. 8th Internat. Congr. Botan. II, 278—280.

— 1956: The Hardiness of Plants. Academic Press. New York.

— 1957 a: The moment of frost injury. Protoplasma **48**, 289—302.

— 1957 b: The role of cell sap concentration in frost hardiness. Plant Physiol. **32**, 237—239.

— and G. W. SCARTH, 1936: Frost-Hardening studies with living cells. Canad. J. Res. **C 14**, 267—305.

LIVINGSTON, J. E., and J. C. SWINBANK, 1950: Some factors influencing the injury to winter wheat heads by low temperatures. Agron. J. **42**, 153—157.

LONA, F., 1956: Recenti esperienze sulle proprietà dell'acido gibberellico. Riv. internat. d'agricol. **10—11.**

— A. SQUARZA, A. BOCCHI e G. CANTONI, 1956: Le reazione al freddo di piante erbacee in rapporto all'azione di sostanze stimolanti ed inibenti l'attività plasmatica. Pub. Chim. Biol. e Med. **2**, 473—494.

LOVELOCK, J. E., 1953 a: The haemolysis of human red blood-cells by freezing and thawing. Biochim. Biophys. Acta **10**, 414—426.
— 1953 b: The mechanism of the protective action of glycerol against haemolysis by freezing and thawing. Biochim. Biophys. Acta **11**, 28—36.
— 1954 a: Biophysical aspects of the freezing and thawing of living cells. Proc. Roy. Soc. Med. **47**, 60—62.
— 1954 b: The protective action of neutral solutes against haemolysis by freezing and thawing. Biochem. J. (Brit.) **56**, 265—270.
— and C. POLGE, 1954: The immobilization of spermatozoa by freezing and thawing and the protective action of glycerol. Biochem. J. (Brit.) **58**, 618—622.
LUCAS, J. W., 1954: Subcooling and ice nucleation in lemons. Plant Physiol. **29**, 245—251.
LUYET, B. J., 1954: Le mécanisme du gel et la résistance au froid. 8th Internat. Congr. Bot. **11**, 259—267.
— and P. M. GEHENIO, 1938: The lower limit of vital temperatures. Biodynam. **33**, 1—92.
MARRÉ, E., and O. SERVETTAZ, 1956: Ricerche sull'adattamento proteico in organismi termoresistente. I. Sul limite di resistenza all'inattivazione termica dei sistemi fotosintetico e respiratorio di alghe di acque termali. Accad. Naz. d. Lincei. Rend. della Classe d. Sci. Fis., Mat. e Nat. Ser. VIII, **20**, 72—77.
— — 1957: Ricerche sull'adattamento proteico in organismi termoresistente. II. Sulla termoresistenza «in vitro» del sistema citocromo riduttasico di Cianoficee termali. Red. dell'Accad. Naz. d. Lincei Cl. Sci. fis., mat. e Nat. Ser. **8**, 22, 91—98.
MAXIMOV, N. A., 1929: The Plant in Relation to Water. Allen and Unwin, London.
— 1931: The physiological significance of the xeromorphic structure of plants. J. Ecol. **19**, 273—282.
MAZUR, P., 1956: Studies on the effects of subzero temperatures on the viability of spores of *Aspergillus flavus*. I. The effect of rate of warming. J. Gen. Physiol. (Am.) **39**, 869—888.
— M. A. RHIAN, and B. G. MAHLANDT, 1957: Survival of *Pasteurella Tulariensis* in sugar solutions after cooling and warming at subzero temperatures. J. Bacter. (Am.) **73**, 394—397.
MERYMAN, H. T., 1956: Mechanics of freezing in living cells and tissues. Science **124**, 515—521.
MILLER, E. C., 1931: Plant Physiology. McGraw-Hill, New York.
MILLER, S. R., and W. G. CORNS, 1957: The constitution of sugar beet seedlings associated with chemically induced improvement in their low temperature resistance. Canad. J. Bot. **35**, 5—8.
MODLIBOWSKA, IRENA, and W. S. ROGERS, 1955: Freezing of plant tissues under the microscope. J. Exper. Bot. 6. 384—391.
— and J. P. RUXTON, 1954: The effect of maleic hydrazide on the spring frost resistance of the Malling Exploit raspberry. J. Hort. Sci. **29**, 184—191.
MORETTI, A., 1953: Physiological effects of winter treatments of chemicals upon grape-vine. Riv. Frutticolt. **15**, 2—25 (Biol. Abstr. **27**, 26044, 1953).
MUELLER-STOLL, W. R., 1947: Der Einfluß der Ernährung auf die Xeromorphie der Hochmoorpflanzen. Planta **35**, 225—251.
OCHI, H., 1952: Autecological study of mosses in respect to water economy. I. On the minimum hydrability within which mosses are able to survive. Bot. Mag. **65**, 112—118.
— 1952 a: On the relationship between so-called "xeromorphic" characters found in mosses and their drought resistance. Bull. Soc. Plant Ecol.. **1**, 182—187.
— 1952 b: The preliminary report on the osmotic value, permeability, drought and old resistance of mosses. Bot. Mag. **65**, 10—12.
OPPENHEIMER, H. R., 1932: Zur Kenntnis der hochsommerlichen Wasserbilanz mediterraner Gehölze. Ber. dtsch. bot. Ges. **50 A**, 185—245.
— 1949: The water turn-over of the *Valonea* Oak. Palestine J. Bot. (Rehovot Ser.) **7**, 177—179.
— 1951: Summer drought and water balance of plants growing in the Near East. J. Ecol. **39**, 356—362.
— 1953: An experimental study on ecological relationships and water expenses of Mediterranean forest vegetation. Palestine J. Bot. (Rehovot Ser.) **8**, 103—124.
ORDIN, L., T. H. APPLEWHITE, and J. BONNER, 1956: Auxin-induced water uptake by *Avena* coleoptile sections. Plant Physiol. **31**, 44—53.

Parker, J., 1951: Moisture retention in leaves of conifers of the Northern Rocky Mountains. Bot. Gaz. **113**, 210—216.

— 1954: Available water in stems of some Rocky Mountain conifers. Bot. Gaz. **115**, 380—385.

— 1955: Annual trends in cold hardiness in *ponderosa* pine and grand fir. Ecology **36**, 377—380.

— 1956: Drought resistance in woody plants. Bot. Rev. **22**, 241—289.

Petinov, N. S., and G. A. Zak, 1938: Effect of hardening upon structure of plant. C. r. (Doklady) Acad. Sci. U. S. S. R. **19**, 543—547.

Petrie, A. H. K., and J. I. Arthur, 1943: Physiological Ontogeny in the Tobacco Plant. The effect of varying water supply on the drifts in dry weight and leaf area and on various components of the leaves. Austral. J. Exper. Biol. a. Med. Sci. **21**, 191—200.

Pieniazek, J., and J. Wisniewska, 1954: The properties of the protoplasm in the tissues of one-year old fruit-tree shoots in the course of different phenophases. Bul. Acad. Pol. Sci. Cl. II **2**, 149—152.

Pirson, A., and E. Goellner, 1953: Beobachtungen zur Entwicklungsphysiologie der *Lemna minor* L. Flora **140**, 485—498.

Pisek, A., and E. Winkler, 1953: Die Schließbewegung der Stomata bei ökologisch verschiedenen Pflanzentypen in Abhängigkeit vom Wassersättigungszustand der Blätter und vom Licht. Planta **42**, 253—278.

Primault, B., 1954: L'influence de l'insolation sur la température du cambium des arbres fruitiers. Rév. Romande Agric. Viticult. et Arboricult. **10**, 26—28.

Rachie, K. O., and A. R. Schmid, 1955: Winter hardiness of birdsfoot trefoil strains and varieties. Agron. J. **47**, 155—157.

Raheja, P. C., 1951: Recent physiological investigations on drought resistance in crop plants. Indian J. Agr. Sci. **21**, 335—346.

Reynolds, E. S., 1939: Tree temperatures and thermostasy. Ann. Mo. Bot. Gard. **26**, 165—255.

Ried, A., 1953: Photosynthese und Atmung bei xerostabilen und xerolabilen Krustenflechten in der Nachwirkung vorausgegangener Entquellungen. Planta **41**, 436—438.

Rodger, J. B. A., C. C. Williams, and R. L. Davis, 1957: A rapid method for determining winter hardiness in alfalfa. Agron. J. **49**, 88—92.

Rogers, W. S., Irena Modlibowska, J. P. Ruxton, and C. H. W. Slater, 1954: Low temperature injury to fruit blossom. IV. Further experiments on water-sprinkling as an anti-frost measure. J. Hort. Sci. **29**, 126—141.

Ruelke, O. Ch., and D. Smith, 1956: Overwintering trends of cold resistance and carbohydrates in medium red, ladino and common white clover. Plant Physiol. **31**, 364—368.

Sakai, A., 1955 a: The relationship between the process of development and the frost hardiness of the mulberry tree. Low Temp. Sci. Ser. B, **13**, 21—31.

— 1955 b: The seasonal changes of the hardiness and the physiological state of the cortical parenchyma cells of mulberry tree. Low Temp. Sci. Ser. B, **13**, 33—41.

— 1956 a: The effect of temperature on the maintenance of the frost hardiness. Low Temp. Sci. Ser. B, **14**, 1—6.

— 1956 b: The effect of temperature on the hardening of plants. Low Temp. Sci. Ser. B, **14**, 7—15.

— 1956 c: Survival of plant tissue at super low temperatures. Low Temp. Sci. Ser. B, **14**, 17—23.

— 1957: The effect of maleic hydrazide upon the frost hardiness of twig of mulberry tree. J. Sericultural Sci. Japan **26**, 13—20.

Salt, R. W., 1950: Time as a factor in the freezing of undercooled insects. Canad. J. Res. D, **28**, 285—291.

— 1953: The influence of food on coldhardiness of insects. Canad. Entomol. **85**, 261—269.

— 1955: Extent of ice formation in frozen tissues and a new method for its measurement. Canad. J. Zool. **33**, 391—403.

— 1956 a: Freezing and melting points of insect tissues. Canad. J. Zool. **34**, 1—5.

— 1956 b: Influence of moisture content and temperature on coldhardiness of hibernating insects. Canad. J. Zool. **34**, 283—294.

Sapper, I., 1935: Versuche zur Hitzeresistenz der Pflanzen. Planta **23**, 518—556.

SATO, Y., et al. 1952: Factors affecting cold resistance in tree seedlings: I. On the effect of period of photoperiodic treatment. II. On the effect of potassium salts. Res. Bull. Coll. Exper. Forests 15, 63—96 (Biol. Abstr. 26, 32259, 1952).

SATOO, T., 1956: Drought resistance of some conifers at the first summer after their emergence. Bul. Tok. Univ. For. 51, 1—108.

SCARTH, G. W., and J. LEVITT, 1937: The frost-hardening mechanism of plant cells. Plant Physiol. 12, 51—78.

SCHIEFERSTEIN, R. H., and W. E. LOOMIS, 1956: Wax deposits on leaf surfaces. Plant Physiol. 31, 240—247.

SCHOLANDER, P. F., W. FLAGG, R. J. HOCK. and L. IRVING, 1953: Studies on the physiology of frozen plants and animals in the Arctic. J. cellul. a. comp. Physiol. (Am.) 42, 1—56.

SCHOPMEYER, C. S., 1939: Transpiration and physico-chemical properties of leaves as related to drought resistance in loblolly pine and shortleaf pine. Plant Physiol. 14, 447—462.

SCHRATZ, E., 1931: Zum Vergleich der Transpiration xeromorpher und mesomorpher Pflanzen. J. Ecol. 19, 292—296.

SEIFRIZ, W., 1955: The physical chemistry of cytoplasm. Handbuch der Pflanzenphysiologie 1, 340—382.

SHARPE, R. H., G. H. BLACKMANN, and Nathan GAMMON, Jr., 1954: Relation of potash and phosphate to cold injury of Moore pecans. Better Crops with Plant Food 38, 17—18.

SHARPLES, G. C., and L. BURKHARDT, 1954: Seasonal changes in carbohydrates in the Marsh grapefruit tree of Arizona. Proc. Amer. Soc. Hort. Sci. 63, 74—80.

SHAW, R. H., 1954: Leaf and air temperatures under freezing conditions. Plant Physiol. 29, 102—104.

SHEAR, C. B., 1953: Zinc in relation to cold injury. Proc. Amer. Soc. Hort. Sci. 61, 63—67.

SHIELDS, LORA MANGUM, 1950: Leaf xeromorphy as related to physiological and structural influences. Bot. Rev. 16, 399—446.

— 1951: The involution mechanism in leaves of certain xeric grasses. Phytomorphology 1, 225—241.

SHMUELI, E., 1953: Irrigation studies in the Jordan Valley. I. Physiological activity of the banana in relation to soil moisture. Bull. Res. Counc. Israel 3, 228—247.

SIMINOVITCH, D., and J. LEVITT, 1941: The relation between frost resistance and the physical state of protoplasm. II. Canad. J. Res. C 19, 9—20.

SIMONIS, W., 1947: CO_2-Assimilation und Stoffproduktion trocken gezogener Pflanzen. Planta 35, 188—224.

— 1952: Untersuchungen zum Dürreeffekt. I. Morphologische Struktur, Wasserhaushalt, Atmung und Photosynthese feucht und trocken gezogener Pflanzen. Planta 40, 313—332.

SLATYER, R. U., 1955: Studies of the water relations of crop plants grown under natural rainfall in northern Australia. Austral. J. Agr. Res. 6, 365—377.

SMITH, A. U., C. POLGE, and J. SMILES, 1951: Microscopic observation of living cells during freezing and thawing. J. Roy. Microsc. Soc. 71, 186—195.

— 1954: The use of glycerol for preservation of living cells at low temperatures. Proc. Soc. Med., Lond. 47, 57—60.

SMITH, D., 1949: Differential survival of ladino and common white clover encased in ice. Agron. J. 41, 230—234.

— 1952: The survival of winter-hardened legumes encased in ice. Agron. J. 44, 469—473.

— 1955 a: Underground development of alfalfa crowns. Agron. J. 47, 588—589.

— 1955 b: Influence of area of seed production on the performance of Ranger alfalfa. Agron. J. 47, 201—205.

— 1957: Flowering response and winter survival in seedling stands of medium red clover. Agron. J. 49, 126—129.

SPRAGUE, V. G., and L. F. GRABER, 1940: Physiological factors operative in ice sheet injury of alfalfa. Plant Physiol. 15, 661—673.

— — 1943: Ice sheet injury to alfalfa. J. Amer. Soc. Agron. 35, 881—894.

— 1955: The influence of rate of cooling and winter cover on the winter survival of ladino clover and alfalfa. Plant Physiol. 30, 447—451.

STEUBING, LORE, 1955: Studien über den Taufall als Vegetationsfaktor. Ber. dtsch. bot. Ges. 68, 55—70.

STOCKER, O., 1929: Eine Feldmethode zur Bestimmung der momentanen Transpirations- und Evaporationsgröße. Ber. dtsch. bot. Ges. **47**, 130—136.

— 1956: Die Dürreresistenz. Handbuch der Pflanzenphysiologie **3**, 696—741.

STONE, E. C., A. Y. SHACHORI, and R. G. STANLEY, 1956: Water absorption by needles of *Ponderosa* pine seedlings and its internal redistribution. Plant Physiol. **31**, 120—126.

SZIRMAI, J., 1938: Die Dörrfleckenkrankheit (Hitzeschaden) des Paprikas. Phytopath. Z. **11**, 1—13.

TABOR, P., 1951: Tolerance of young seedlings of *Lespedeza* and *Kudzu* to cold and naphtha. Agron. **43**, 205.

TENNANT, J. R., 1954: Some preliminary observations on the water relations of a mesophytic moss, *Thamnium alopecurum* (Hedw.) B. and S. Trans. Brit. Bryol. Soc. **2**, 439—445.

THODAY, D., 1931: The significance of reduction in the size of leaves. J. Ecol. **19**, 297—303.

THOMAS, M., 1951: Carbon dioxide fixation and acid synthesis in Crassulacean acid metabolism. Symp. Soc. Exper. Biol. **5**, 72—93.

THORNTHWAITE, C. W., 1956: The air as a water absorbing medium. Handbuch der Pflanzenphysiologie **3**, 257—264.

TILL, O., 1956: Über die Frosthärte von Pflanzen sommergrüner Laubwälder. Flora **143**, 499—542.

TODD, G. W., and J. LEVITT, 1951: Bound water in *Aspergillus niger*. Plant Physiol. **26**, 331—336.

TONZIG, S., 1941: I Muco-proteidi e la vita della cellula vegetale. Libreria Universitaria di G. Randi. Padova.

TORSSELL, B., and N. HELLSTRÖM, 1955: Investigations on oil turnips and oil rape. IV. Estimation of plant status. Acta Agric. Scand. **5**, 31—38.

VAN BAVEL, C. H. M., 1953: A drought criterion and its application in evaluating drought incidence and hazard. Agron. J. **45**, 167—172.

VEIHMEYER, F. J., 1956: Soil moisture. Handbuch der Pflanzenphysiologie 3, 64—123.

VICKERY, H. B., 1953: The behavior of organic acids and starch of *Bryophyllum* leaves during culture in continuous light. J. Biol. Chem. (Am.) **205**, 369—381.

WALTER, H., 1950: Einführung in die Phytologie I. Die Grundlagen des Pflanzenlebens 72—78, 326—356.

— 1955: The water economy and the hydrature of plants. Ann. Rev. Plant. Physiol. **6**, 239—252.

— 1956: Die heutige ökologische Problemstellung und der Wettbewerb zwischen der mediterranen Hartlaubvegetation und den sommergrünen Laubwäldern. Ber. dtsch. bot. Ges. **69**, 263—273.

WAY, R. D., 1954: The effect of some cultural practices and of size of crop on the subsequent winter hardiness of apple trees. Proc. Amer. Soc. Hort. Sci. **63**, 163—166.

WERK, O., 1954: Untersuchungen zum Dürreeffekt. 2. Über den Kalium- und Calciumgehalt feucht und trocken gezogener Pflanzen. Flora **141**, 312—355.

WILNER, J., 1955 a: Results of laboratory tests for winter hardiness of woody plants by electrolytic methods. Proc. Amer. Soc. Hort. Sci. **66**, 93—99.

— 1955 b: The effect of low temperatures on available soil moisture during winters on the Canadian Prairies. Agron. J. **47**, 411—413.

WOOD, R. R., H. E. BREWBAKER, and H. L. BUSH, 1950: Studies of cold resistance in sugar beets. Proc. Amer. Soc. Sugar Beet Technol. 116—121.

— and M. A. SPRAGUE, 1952: Relation of organic food reserves to cold hardiness of Ladino clover (*Trifolium repens* L.). Agron. J. **44**, 318—325.

YA-E, M., 1955: Study on the Plasmolysis in Epidermal cells from leaves of *Saxifraga stolonifera* Meerb. Mem. Fac. Educ. Kumamoto Univ. **3**, 163—176.

Index